AF819133

SOTHIS PRESS

Sothis Press aims to completely reimagine the public domain book with an emphasis on universal access to captivatingly designed print editions. We believe that technology will enable every single book ever published to be perpetually accessible in beautiful print form. Sothis is beauty in ubiquity.

www.sothis.press
1 Waterhouse Square, London, ECIN 2ST

A Tract

On the Possible and Impossible Cases of Quadratic Duplicate Equalities in the Diophatine Anylysis

Matthew Collins

SOTHIS PRESS

ISBN 978-1-0368-0114-4

This edition is published by
Sothis Press © 2023. The company
is registered in England and Wales.
The Company Number is 14900389.

A TRACT

ON THE POSSIBLE AND IMPOSSIBLE CASES OF QUADRATIC DUPLICATE EQUALITIES IN THE DIOPHANTINE ANALYSIS:

TO WHICH IS ADDED

A SHORT, BUT COMPREHENSIVE APPENDIX,

IN WHICH MOST OF THE USEFUL AND IMPORTANT

Propositions in the *Theory of Numbers* are very concisely demonstrated.

BY MATTHEW COLLINS, B.A.
Senior Moderator in Mathematics and Physics, and Bishop Law's Mathematical Prizeman, Trin. Coll. Dublin.

Nil tam difficile est quod non Solertia Vincat.

Pythagoras regarded Arithmetic as the noblest Science, and an acquaintance with Numbers as the highest good.

DUBLIN:
This Tract and Mr. Collins' other works, can be procured direct from the Author, 4 Ormonde-road, Kilkenny, or, 13 Anglesea-street, Dublin.
They are sold by
MESSRS. PONSONBY, CORNISH, KELLY, ROONEY, and MORRIS of DUBLIN: MULCAHY, CORK: O'GORMAN, LIMERICK: DOUGLAS and NICHOLSON, KILKENNY.
1858.

[*Price Four Shillings.*]

DUBLIN:

PRINTED BY GOODWIN, SON, AND NETHERCOTT, 79, MARLBOROUGH-ST.

To

WILLIAM SMITH O'BRIEN, ESQ.,

THE NOBLE, PATRIOTIC, AND LIBERAL PATRON OF SCIENCE,

LITERATURE, AND IRISH TALENT,

IN GRATEFUL ACKNOWLEDGMENT OF MANY KIND

FAVORS RECEIVED FROM HIM,

THIS TRACT IS RESPECTFULLY DEDICATED,

BY HIS MOST OBLIGED,

HUMBLE SERVANT,

THE AUTHOR.

PREFACE.

About three years ago I first began to consider and meditate on the subject matter of the following Tract, solely for the delightful pleasure and amusement the pursuit afforded me, and without the least idea of publication. By slow degrees I fell upon a systematic and regular method of proving the *impossibility* of several of the duplicate equalities treated of in the first two chapters of the Work, and by careful induction from the particular cases, I was soon led to the discovery of the *new and useful general theorems* for solving the *possible cases* given in Articles 10, 19, and 36, which enable us to find a set of answers (to a quadratic duplicate equality), in greater integers from a given set in less integers, by *a method much more direct and expeditious than Fermat's.* Indeed I believe these theorems are the first and only instance yet found for dispensing at all with the use of Fermat's method in such cases, and it is quite surprising that their discovery escaped the acute penetration of the *illustrious Euler* who touched so closely upon them in Art. 230 of his Algebra, which Art. was my first clue, insight, and guide to this delightful province. My discovery of these general theorems, and of those given in Arts. 28 and 41, shews that *Euler* did not deduce from his Art. 230 above-mentioned, all the useful consequences that naturally follow from it.

It is only very recently I perceived a most useful and important application of these general theorems, and other propositions proved in this Tract, in establishing (see pages 30 and 31) the *impossibility* of the equation, $ax^4+bx^2y^2+cy^4=\square$ in a *great* number of cases, and thus making very extensive and valuable additions to the *few* cases of this kind heretofore recorded and proved by Fermat and Euler.

In order to render this work as useful as possible in promoting the advancement of science and education, and to contribute, as far as I could, to spread a knowledge of *the theory of numbers*, throughout the United Kingdom, I have added to it an Appendix containing *the latest and best demonstrations* of several of the most useful and important propositions in that delightful science, a matter which was certainly much needed, as the only English work (Barlow's) on the subject is long since out of print and cannot be procured, and especially as some of the demonstrations have been since *improved and simplified,* and as no good sketch even, of the subject was given in any English work on Algebra.

I have applied *in vain* to the Board of T.C.D., and to the Irish Board of National Education for some little assistance towards defraying the expense of printing, &c. I own I was greatly surprised, disappointed and grieved at finding both these Boards (from whom I *foolishly* expected active assistance) so extremely illiberal and penurious as to refuse giving me the least help. Alas! how really useless to a country are such corporations, who would sooner let discoveries in science perish than advance one shilling to forward their publication or to assist native talent. The reply of the *National* Board to my application was scarcely civil, and disgraced by bad grammar and even by bad spelling; but that institution has been long notorious for totally disregarding *merit* compared to *private interest* in the appointment of its officers.

MATTHEW COLLINS.

13, *Anglesea-street, Dublin.*
August 18*th*, 1858.

A TRACT

ON THE

POSSIBLE AND IMPOSSIBLE CASES OF QUADRATIC DUPLICATE EQUALITIES.

CHAPTER I.

On the possible and impossible cases of $x^2+ay^2=\square$ *and* $x^2-ay^2=\square$

Article 1.—To prove that all the possible solutions in whole numbers prime to each other of the equation $x^2+y^2=z^2$ are contained in the formulæ $x=p^2-q^2$, $y=2pq$ and $z=p^2+q^2$; p and q being prime to each other, one of them even and the other odd.

DEMONSTRATION.

It appears from the given equation itself that if any two of x, y, z, had a common measure, the 3rd should have the same common measure, and then dividing the equation by $(\text{G C M})^2$ we would obtain a similar equation, in which the three variables would be prime to each other, and this being supposed it is evident that x and y cannot be both odd, for if so x^2+y^2 could not be a square number, as it would be even and its half odd; and as x is prime to y they cannot both be even ∴ one of them must be even and the other odd. Let it be y that is even, and ∴ x and z must then be both odd, else they would not be prime to y, we may ∴ take $y=2y'$, $z=z'+x'$ and $x=z'-x'$, z' being prime to x' (else z would not be prime to x), one even and the other odd (else z and x would not be both odd), and then $y^2=z^2-x^2$ gives $y'^2=z'x'$, and here, as z' is prime to x', and their product a square ∴ each factor must be itself a square; let ∴ $z'=p^2$ and $x'=q^2$ and ∴ $y'=pq$, and as z' is prime to x', one even and the other odd ∴ p and q must be prime to each other, one even and the other odd; then $y=2y'=2pq$, $x=z'-x'$ ∴ $=p^2-q^2$ and $z=z'+x'$ ∴ $=p^2+q^2$, and hence these formulæ contain all the possible solutions in integers, prime to each other, of the proposed equation $x^2+y^2=z^2$. Q.E.D.

Cor.—We need scarcely remark that when we have found three rational numbers, x, y, z, such as to fulfil the equation $ax^2+by^2=cz^2$, then mx, my, and mz will be other values of x, y, z, to satisfy the same equation, and these new values will be rational and integral when m and the former values are so, but they will always preserve the proportions of the original x, y, z.

2.—This method is obviously applicable to the solution of the more general equation $z^2-x^2=\text{A}y^2$; first, if A contains a square factor, m^2, this factor can be suppressed, or rather incorporated with y, for if $\text{A}=am^2$, m^2 being the *greatest* square factor in A, then putting y' for my, the proposed equation becomes $z^2-x^2=ay'^2$, in which the coeft. a no longer contains a square factor; and if this last equation be possible at all it must obviously be so when x, y', and z are all prime to each other, for the equation shows that z or x should contain any factor common to the other two, $x\,y'$, or $z\,y'$, it also shows that y' should contain a factor common to z and x since ay'^2 can contain no square factor but those contained in y'^2 as a itself contains no square factor, and thus it appears that if $z^2-x^2=ay'^2$ be possible at all it must be so when x, y', z, are all prime to each other, and it is this *primitive* solution if it exists that we always seek in the present and similar instances.

Now the equation $ay'^2=z^2-x^2=(z+x)(z-x)$ shows that some one of the factors $z\pm x$ must be divisible by a if a be a prime number, but if a be composite and $=bc$, then it is not absolutely necessary that either $z\pm x$ should be divisible by a as the equation will obviously be fulfilled by having one of the two factors $z\pm x$ divisible by b, and the other by c, and as x is prime to z ∴ $z+x$ is prime to $z-x$ or else they can have no common measure but 2, which happens when z and x are both odd. When a is even and $=2a'$ tis evident z and x cannot be even and odd, they must ∴ be both odd, so that we may then put $z=z'+x'$, $x=z'-x'$, x' being prime to z' (else x would not be prime to z) one even and the other odd (else z and x would not be both odd) and then $ay'^2=z^2-x^2$ gives $a'y'^2=2z'x'$. Now if x' be odd and ∴ z' even, then x' will be prime to $2z'$ as it is prime to z' and ∴ the quotients $\frac{2z'}{b'}$ and $\frac{x'}{c'}$ will be prime to each other $(a'=b'c')$ and as their product is a square (y'^2) ∴ each factor must be a square, but since a' is odd, for $2a'=a$ contains no square factor ∴ b' and c' must be both odd, and ∴ $\frac{2z}{b'}$ will be even and $\frac{x'}{c'}$ odd, so that we shall have $\frac{2z'}{b'}=(2p)^2$ and $\frac{x'}{c'}=q^2$, q being odd and prime to p; these give $y'=2pq$, $z=z'+x'$ ∴ $=2b'p^2+c'q^2$ and $x=z'-x'$ ∴ $=2b'p^2-c'q^2$ and $y=\frac{y'}{m}$ ∴ $=\frac{2pq}{m}$: but if z' were odd and ∴ x' even, then we should have $\frac{2x'}{b'}=(2p)^2$ and $\frac{z'}{c'}=q^2$, q as before being odd and prime to p; these give $-x=x'-z'=2b'p^2-c'q^2$, and exactly the same value as before for y and z, thus the sign of x becomes changed, but this is of no consequence since it is x^2 and not x itself that occurs in the proposed equation. Indeed both cases will be included in the general formulæ $\pm\ x=2b'p^2-c'q^2$, $y=2pq$ and $z=2b'p^2+c'q^2$; q being odd and prime to p, and by interchanging b' and c' tis evident we will have just twice as many formulæ or solutions as there are different ways of resolving a' $(=b'c')$ into two factors. When a' is a prime number then one of b' and c' will obviously be $=1$ and the other $=a'$.

Ex. gr. The *primitive* solution of $z^2-x^2=14y^2$ is $y=2pq$ and,

$$\begin{Bmatrix} z=14p^2+q^2 & \text{or} & =2p^2+7q^2 \\ \pm x=14p^2-q^2 & \text{or} & =2p^2-7q^2 \end{Bmatrix}$$

and the *primitive* solutions of $z^2-x^2=270y^2$ where $m=3$ and $b'c'=15$ are $y=\frac{2pq}{3}$ and then

$$\begin{array}{|r|l|l|l} z=30p^2+q^2 & 2p^2+15q^2 & 6p^2+5q^2 & 10p^2+3q^2 \\ \pm x=30p^2-q^2 & 2p^2-15q^2 & 6p^2-5q^2 & 10p^2-3q^2 \end{array}$$

q being odd and prime to p, and having always an odd coeft. in the values of z and x: and these general formulæ contain all the possible solutions in whole numbers of the proposed equation, $z^2-x^2=Ay^2=2b'c'm^2y^2$ if as is directed in Cor. Art 1, we multiply the values of x, y, z here determined by m' or any multiple of m' where m' is a divisor of m and equal to denominator of $\frac{2pq}{m}$ when reduced to its lowest terms.

3—When a is odd and $=b'c'$ then if y' be odd the equation $ay'^2=z^2-x^2=(z+x)(z-x)$ shows that x and z must be one even and the other odd, and ∴ $z\pm x$ both odd and prime to each other, and the equation cannot subsist unless one of the two prime factors $z\pm x$ be divisible by a or else one of them divisible by b' and the other by c', and as the two quotients will be prime to each other, and their product a square (y'^2) ∴ each must be a square, so that we shall have $\frac{z+x}{b'}=p^2$ and $\frac{z-x}{c'}=q^2$, p being prime to q and both odd; these give $z=\frac{1}{2}(b'p^2+c'q^2)$; $\pm x=\frac{1}{2}(b'p^2-c'q^2)$, $y'=pq$ and $\therefore y=\frac{y'}{m}=\frac{pq}{m}$ for the general solution in this case; and as interchanging b' and c' would not at all affect the value of y and z and only change the sign of x, we ∴ include all possible cases by placing the sign $\pm$ before x, and 'tis evident we will thus obtain as many formulae or solutions as there are different ways of resolving $a\ (=b'c')$ into two factors.

But if when a is odd y' be even, then $ay'^2=z^2-x^2$ shows that z and x must be both odd, so that we may then put $z=z'+x'$ and $x=z'-x'$ and $y'=2y''$, x' being prime to z' one even the other odd, then the equation becomes $ay''^2=z'x'$, which cannot subsist. unless one of z', x' be divisible by $a\ (=b'c')$, or else one of them divisible by b' and the other by c', so that we shall have $\frac{z'}{b'}=p^2$ and $\frac{x'}{c'}=q^2$ and $\therefore y''=pq$ and $y'=2pq$, which give $z=z'+x'=b'p^2+c'q^2$, $\pm x=b'p^2-c'q^2$ and $y=\frac{y'}{m}=\frac{2pq}{m}$ which are just double the values of x, y, z, found in the preceding case; so that we must remember that these expressions are to be halved, in order to find the *primitive* values of x, y, z, when p and q are both odd, as well as y'

Ex, gr—the primitive solution of $z^2-x^2=21y^2$ when y is even, is $y=2pq$ and $\left\{\begin{array}{l} z=21p^2+q^2 \text{ or } =7p^2+3q^2 \\ \pm x=21p^2-q^2 \text{ or } =7p^2-3q^2 \end{array}\right\}$ p being prime to q, one even and the other odd; but when y is odd, p and q must be odd, and then we must use the halves of these expressions; for x, y, and z.

4.—If the general equation $ax^2+by^2=cz^2$ be possible when a, b, c, x, y, z, are rational, it must obviously be so too, when these quantities are integers, in the same proportion, as we would obtain a similar equation, by multiplying it by the product of the denominators of a, b, c, x^2, y^2 and z^2, and then we could suppress by division a factor, common to a, b, c, or to x^2, y^2, and z^2; so conversely in seeking the solution of such an equation, the integers, x, y, z, are always sought in this *primitive* state, where the whole three of them have no common measure, the common divisor of a, b, c, being previously suppressed by division; then if one or more of the coefts. contain square factors, they may be rejected, or rather incorporated with x, y, z: thus if $a=a'A^2$, $b=b'B^2$ and $c=c'C^2$, then putting $x'=Ax$ $y'=By$, and $z'=Cz$, the proposed equations, $ax^2+by^2=cz^2$ becomes $a'x'^2+b'y'^2=c'z'^2$, where the coefts. a', b', c', no longer contain any square factors, and so if the original equation were possible, this last equation should be so, and conversely, if this last equation be impossible, the original one must also be impossible; and from the solution of this latter equation, we can return and obtain a solution of the proposed equation, by multiplying this latter equation by $A^2\ B^2\ C^2$, and then taking $x=BCx'$, $y=ACy'$ and $z=ABz'$ and restoring a for $a'A^2$, b for $b'B^2$, and c for $c'C^2$, as is self-evident. But when we have to solve two *simultaneous* equations,

$$\left\{\begin{array}{l} ax^2+by^2=cz^2 \\ a'x^2+b'y=c'w^2 \end{array}\right\}$$

we cannot then suppress any square factors in the coefts., except such as are common to a and a', or to b and b', or to c and c', for if by the above method, we suppressed such factors in one equation, they would re-appear in the other equation.

5—Now I say, if the derived equation $a'x'^2+b'y'^2=c'z'^2$ be possible at all, it must be so, when x' y' z' are *all* prime to each other; for if any two of them x', y' had a common factor m the equation, shows that $c'z'^2$ would have the factor m^2, and as by hypothesis c' contains no square factor $\therefore$ $c'z'^2$ can have no square factor, except those of z'^2, and so m^2 should be a factor of z'^2, and thus every term of the equation being $\div m^2$ we would obtain by division a similar equation, in which x', y', z', would be prime to each other; moreover, if any two coefts. a', b' be prime to each other, the 3rd variable z' must be prime to each of them; for if z' and a' had a common measure m the equation shows that m should be a factor of $b'y'^2$, and as m is primitive to b $\therefore$ it should divide y', but in the sought *primitive* solution z' must be prime to y' as already proved, and $\therefore$ if such a solution exists, z' must be prime to a' and b' when these are prime to each other.

It follows also that if any two of the 3 numbers x' y' z' be prime to each other, the whole three of them must be prime to each other since it is already proved that if x' and y' had a common factor m, z' should have the same factor, and then no two of them would be prime to each other, which is contrary to hypothesis; and this proof evidently holds good even when one coeft. c' contains a square factor, provided the coefts. a', b' of the two numbers x', y' supposed prime to each other contain no square factor: and again 'tis evident that any two (x', y') of the three x' y' z' can have no common factor m unless m^2 be a factor of (c') the coeft. of the third number z'^2, so that when the coeft. (c') of z'^2 contains no square factor, then x' must be prime to y' in the *primitive* solution.

These last observations are of use when we have to solve two simultaneous equations— $\begin{cases} ax^2 + by^2 = cz^2 \\ a'x^2 + b'y^2 = c'w^2 \end{cases}$, where some of the coefts. may contain square factors; from these, by eliminating y or x, we derive two other equations— $\begin{cases} Az^2 + Bw^2 = C\,x^2 \\ A'z^2 + B'w^2 = C'y^2 \end{cases}$ and if none of the coefts. of these four equations contain a square factor, it follows, from what has been already said in this Art. 5, that if the two original equations be simultaneously possible they shall be so too when x, y, z, w are all prime to each other (as the two original equations could be rederived from any two of the four equations), and then, too, x must be prime to b, c, b', c', and y must be prime to a, c, a' and c' &c., supposing b prime to c, and b' to c', &c., &c. it follows also that if the two original equations be possible at all, they must be so too when x is prime to y, if neither c nor c' contain a square factor, or if $c=c'=1$, which is the most usual case; and, lastly, it follows, from what has been said, that if $ax^2+by^2=cz^2$ be possible, it must be so, too, when x and y are prime to c if a and b be prime to c, or if $a=\pm1=b$.

6.—The two simultaneous equations, $\begin{cases} x^2+y^2=\square=z^2 \\ x^2-y^2=\square=w^2 \end{cases}$ are impossible.

For by addition and subtraction $2x^2=z^2+w^2$ and $2y^2=z^2-w^2$ and $\therefore$ by the foregoing Art. 5, if the proposed equations be possible they must be so when x, y, z, and w are all prime to each other, and also z and w prime to the coeft. 2 and $\therefore$ both odd, so that we may put $z=z'+w'$ and $w=z'-w'$; z' and w' being prime to each other, one even and the other odd, as shown in Art. 1, then by substitution $z'^2+w'^2=x^2$ and $2z'w'=y^2$ and by Art. 1 the general solution of the first of these two equations is $z'=m^2-n^2$ and $w'=2mn$, m being prime to n, odd and even; these substituted in the second equation give $mn(m+n)(m-n)=\square\left(=\frac{y^2}{4}\right)$ And as the four factors are prime to each other and one of them (m or n) even $\therefore$ each factor must be a square, let $\therefore$ $m=p^2$, $n=q^2$, $m+n=r^2$ and $m-n=s^2$, hence $p^2+q^2=r^2$ and $p^2-q^2=s^2$ a pair of equations precisely similar to the original pair, having p and q instead of x and y; and since in the equation $z'^2+w'^2=x^2$ we have $x=m^2+n^2\therefore=p^4+q^4$ and $y^2=4\,mn\,(m+n)\,(m-n)=4p^2q^2r^2s^2$ and $\therefore y=2pqrs$ $\therefore$ $x>p$ and $y>q$ so that if the proposed equations admit of a solution in great numbers x and y they should also admit of a solution in smaller integers p aud q, and by the same argument there should exist a solution in integers still smaller than p and q, &c., &c., and as there exists no solution in small integers, so neither can there exist any solution even among the largest whole numbers. Q.E.D.

7.—The two simultaneous equations $\left\{\begin{matrix} x^2+2y^2=\square=z^2 \\ x^2-2y^2=\square=w^2 \end{matrix}\right\}$ are impossible.

For if these equations were possible, they should obviously be so when x is prime to y and $\therefore$ all four x, y, z, w, prime to each other except perhaps z and w; and, moreover, x, z and w odd, as being prime to coeft. 2 (of $2y^2$); now by addition and subtraction $z^2+w^2=2x^2$ and $z^2-w^2=4y^2$ and here as this last coeft. 4 is a square number $\therefore$ by Art. 5, z and w could have no factor not common to x and y unless 2 $(=\sqrt{4})$ but as z and w were already proved to be odd, they could not have 2 as common factor, and so if the proposed equations be possible at all they should be so when x, y, z and w are all prime to each other, and all odd except y, so that we may, as in the foregoing Art., put $z=z'+w'$ and $w=z'-w'$, z' and w' being prime to each other, one even and the other odd; substituting we get $z'^2+w'^2=x^2$ and $z'w'=y^2$ and by Art. 1 the general solution of the former is $z'=m^2-n^2$ and $w'=2\,mn$ and $x=m^2+n^2$, m being prime to n one even and the other odd, and $\therefore$ m, n and $m\pm n$ prime to each other and only one of them (m or n) even; these substituted in the second equation give $mn(m+n)(m-n)=\frac{y^2}{2}\therefore=2y'^2$, y being $=2y'$. 1° if m be even we shall obviously have $m=2p^2$, $n=q^2$, $m+n=r^2$ and $m-n=s^2$ and $\therefore 2p^2+q^2=r^2$ and $2p^2-q^2=s^2$; p, q, r and s being prime to each other and the three latter odd, and as every odd square is of the form $8\,n+1$ the equation $2p^2=r^2-q^2$ shows that p must be even, and then $2p^2-q^2$ is of the form $8\,n+7$ which is an *impossible* form for the odd square s^2. 2° If n be even then $m=p^2$, $n=2q^2$, $m+n=r^2$ and $m-n=s^2$ and $\therefore p^2+2q^2=r^2$ and $p^2-2q^2=s^2$ two equations precisely similar to the original pair, only having p and q instead of x and y, but since $x=m^2+n^2\therefore=p^4+4q^4$ and y^2 was $=z'w'\therefore=2mn\,(m+n)\,(m-n)=4p^2q^2r^2s^2$ and $\therefore y=2\,p\,q\,r\,s$, thus $x>p$ and $y>q$ and hence the two proposed equations are impossible for reasons like those already assigned towards the end of the foregoing Article 6.

8.—The two simultaneous equations $\left\{\begin{matrix} x^2+3y^2=\square=z^2 \\ x^2-3y^2=\square=w^2 \end{matrix}\right\}$ are impossible.

By addition and subtraction $2x^2=z^2+w^2$ and $6y^2=z^2-w^2$ and as none of the coefts. in these four equations contains a square factor $\therefore$ by Art. 5, if the proposed equations be possible at all they must be so when x y z and w are all prime to each other and also z and w prime to 6 (in the equation $6y^2=z^2-w^2$) and $\therefore$ both odd, we may $\therefore$ as bofore put $z=z'+w'$ and $w=z'-w'$, z' and w' being prime to each other, one even and the other odd, this substitution gives

$z'^2+w'^2=x^2$ and $z'w'=\frac{3y^2}{2}$ this last shows that y must be even, let $\therefore y=2y'$, now by Art. 1 the general solution of the former is $z'=m^2-n^2$ $w'=2mn$ and $x=m^2+n^2$, m being prime to n, one even and the other odd, and $\therefore m, n, m+n$ and $m-n$ all prime to each other, and one of them (m or n) even; the second equation $z'w'=\frac{3y^2}{2}\therefore=6y'^2$ by this substitution becomes $mn(m+n)(m-n)=3y'^2$ which shows that some one (and only one) of the four prime factors, $m, n, m\pm n$ must be divisible by 3, and then of course the quotient will be prime to each of the other three factors, and as the product of the quote and the other three factors is a square (y'^2) $\therefore$ each factor must be itself a square; 1° if m be divisible by 3 then $m=3p^2$, $n=q^2$, $m+n=r^2$ and $m-n=s^2$ giving $q^2+s^2=3p^2$ an impossible equation, for as q is prime to s they cannot both be divisible by 3, and according as one or none of them is divisible by 3, q^2+s^2 will be of the form $3A+1$ or $3A+2$, and neither of these could $=3p^2$. 2° If n be $\div$ 3 then $m=p^2$, $n=3q^2$, $m+n=r^2$ and $m-n=s^2$ giving $p^2+3q^2=r^2$ and and $p^2-3q^2=s^2$ which are a pair of equations exactly similar to the original pair only having p and q instead of x and y, and as p and q are $<x$ and y $\therefore$ the proposed equations are impossible, for reasons already assigned in Art. 6. 3° if $m+n$ were $\div$ 3 we should have $m=p^2$, $n=q^2$, $m+n=3r^2$ and $\therefore p^2+q^2=3r^2$ which was proved impossible in case 1°. 4° If $m-n$ were $\div$ 3 then $m=p^2$, $n=q^2$ $m+n=r^2$ and $m-n=3s^2$ which give $2p^2=r^2+3s^2$ or $p^2-s^2=\frac{p^2+r^2}{3}$. An equation obviously impossible, for by case 1° when p is prime to r, p^2+r^2 is not divisible by 3 and $\therefore \frac{p^2+r^2}{3}$ could not give the integral quotient p^2-s^2.

Cor. 1 The two simultaneous equations $x^2+2y^2=\square=z^2$ and $y^2+2x^2=\square=w^2$ are impossible. For if they were possible they should by Art. 5 be so when x y z and w are all prime to each other and $\therefore$ when z is prime to w; but they give $x^2+y^2=\frac{z^2+w^2}{3}$ which was proved impossible in the foregoing case I°.

Cor 2. The two expressions $x^2\pm ay^2$ cannot both be squares whenever $a=A^2$ or $=2A^2$ or $=3A^2$, A being any rational number whole or broken as appears instantly from Articles 6, 7, 8 by putting y' for Ay, and hence the two expressions $x^2\pm ay^2$ cannot both be squares when $a=4$, 8, 9, 12, 16, 18, 25, 27, 32, &c.

9. The two simultaneous equations $\left\{\begin{matrix} x^2+5y^2=\square=z^2 \\ x^2-5y^2=\square=w^2 \end{matrix}\right\}$ are possible.

By addition and subtraction $2x^2=z^2+w^2$ and $10y^2=z^2-w^2$, and as none of the coefts. of these four equations contains a square factor $\therefore$ by Art, 5 If the proposed equations be possible they must be so when x, y, z, and w are all prime to each other, and z and w, moreover, both odd, as appears from the equation $2x^2=z^2+w^2$ we may $\therefore$ as usual put $z=z'+w'$, and $w=z'-w'$, z' and w' being as usual prime to each other, one even and the other odd, this done we get $z'^2+w'^2=x^2$, and $z'w'=\frac{5y^2}{2}$. Now this last shows that y must be even, let it $\therefore=2y'$; by Art. 1 the general solution of the former is $z'=m^2-n^2$, $w'=2mn$, $x=m^2+n^2$, which give $mn(m+n)(m-n)=5y'^2$, m and n being as usual, one even and the other odd, and prime to each other, and $\therefore m, n, m \pm n$, all prime to each other. Now this last equation shows that some one of the four prime factors must be divisible by 5 and then the quote will still be prime to the other 3 factors: but the product of any set of factors prime to each other cannot be a square unless each factor be itself a square. 1° if m be $\div$ 5 we shall have $m=5p^2$ $n=q^2$, $m+n=r^2$ and $m-n=s^2$ giving $5p^2+q^2=r^2$ end $5p^2-q^2=s^2$ (and $2q^2=r^2-s^2$) and by Art. 5, if these be resolvible r and s must be prime to 2 (coeft. of $2q^2$) and $\therefore$ both odd, and also q, r, and s must be prime to 5 (coeft. of $5p^2$) and in fact tis evident $p=1$, $q=2$, $r=3$, $s=1$ fulfil these two equations, and then $x=m^2+n^2 \therefore =\frac{1}{2}(r^4+s^4) \therefore =41$ and $y'^2=p^2\,q^2\,r^2\,s^2$ and then $y=2y'=2pqrs$ here $=$ 12 see Euler's Algebra, page 426. 2°.—If n be divisible by 5 we get $m=p^2$. $n=5q^2$, $m+n=r^2$ and $m\ n=s^2$ giving $p^2+5q^2=r^2$ and $p^2-5q^2=s^2$ a pair of equations exactly similar to the original pair, but having p and $q < x$ and y since $x=m^2+n^2 \therefore =p^4+25q^4=\frac{1}{2}(r^4+s^4)$ is $> p$ and $y=2\,pqrs$ is $> q \therefore$ we could not descend to this case (of n divisible by 5) *from the smallest whole numbers* x and y that answer the proposed equations.

But on the other hand the preceding formulae enable us to ascend from one known solution of the proposed equations, to another solution in larger integers; for as $p^2+5q^2=r^2$ and $p^2-5q^2=s^2$ are similar to the proposed equations, and as we found $x=p^4+25q^4=\frac{1}{2}(r^4+s^4)$ and $y=2pqrs$ it is evident that new $X=x^4+25y^4=\frac{1}{2}(z^4+w^4)$ and new $Y=2xyzw$ will also answer for x and y in the proposed equations, from which it is evident we can again find, by these same formulae, new values of x and y, in still larger integers, &c., &c. Thus in the preceding case we found $x=41$ and $y=12$ giving $z=49$ and $w=31$ and hence new $x=\frac{1}{2}(49^4+31^4)=3344161$ and new $y=41\times 24\times 49\times 31=1494696$, from which we could again find new and very great values of x and y, and thus ascend into very high whole numbers, &c., &c. 3°, $m+n$ cannot be divisible by 5, for if it were we should have $m=p^2$, $n=q^2$, $m+n=5r^2$, and $m-n=s^2$ giving $5r^2+s^2=2p^2$ or $r^2=\frac{2p^2-s^2}{5}$, an impossible equation; for as $m+n$ is divisible by $5 \therefore p,q$ and s cannot be so, as they are prime to $m+n$ and $\therefore p^2$, q^2 and s^2 are each of the form $5A\pm 1$ and $\therefore 2p^2-s^2$ could not be divisible by 5, so as to give the integral quotient r^2. 4° neither can $m-n$ be divisible by 5, for if so we should have $m=p^2$, $n=q^2$, $m+n=r^2$, and $m-n=5s^2$, giving $5s^2+r^2=2p^2$, which was just proved impossible in case 3°.

Remark—Thus as neither $m\pm n$ can be divisible by 5, and as already observed, we cannot descend to the case of n divisible by 5, *from the least whole numbers*, x and y, that fulfil the proposed equations; and moreover as p and q are $< x$ and y, it follows $\therefore$ that the equations, $5p^2+q^2=r^2$ and $5p^2-q^2=s^2$ (or $x^2+5y^2=z^2$ and $x^2-5y^2=-w^2$) admit of a solution in integers, less than even the least integers that answer the proposed equations; this remark is important on several occassions, Ex gr it applies well to the equations, $x^2\pm 13y^2=\square^s$, see cor. 3, Art. 14.

10—The Invertigations in the preceding Art., 9, may be rendered more general thus:—

General Theorem.—The solution of $X^2+ab\,Y^2=\square=Z^2$ and $X^2-aby^2=\square=W^2$ can be obtained from a solution of the two auxiliary equations, $ax^2+by^2=nz^2$ and $ax^2-by^2=\pm nw^2$, in fact I say $X=\frac{1}{2}n(z^4+w^4)$ and $Y=2xyzw$ will answer.

Demonstration.—The difference of the squares of the two auxiliary equations gives $4abx^2y^2=n^2(z^4-w^4)$ and $abY^2=4abx^2y^2z^2w^2 \therefore =n^2z^2w^2(z^4-w^4)$ and as $4X^2=n^2(z^4+w^4)^2=n^2(z^4-w^4)^2+n^2(2z^2w^2)^2=n^2(t^2+v^2)$ where $t=z^4-$

w^4 and $v=2z^2w^2$ and $4abY^2$ is $=4n^2z^2w^2(z^4-w^4)$.$\therefore=n^2\times 2tv$.$\therefore 4(X^2\pm abY^2)$ $=n^2(t\pm v)^2$, which are both squares, Q. E. D.

The generality attained here by the use of the indeterminate quantity n applies only to the auxiliary equations, since $Y^2=\dfrac{n^2z^2w^2(z^4-w^4)}{ab}$ shows that Y^2, as well as X^2 contains the factor n^2, which ought $\therefore$ to be rejected from both; the use of n serves, however, to prove certain classes of quadratic duplicate equalities impossible or incompatible; *Ex, gr;* as $x^2+y^2=z^2$ and $x^2-y^2=w^2$ were proved impossible in Art. 6; hence we infer that $x^2+y^2=nz^2$ and $x^2-y^2=nw^2$ must be also impossible, whatever rational number n may be; for if these latter were possble, the former should be so by this present Art. 10; in like manner as $x^2+2y^2=z^2$ and $x^2-2y^2=w^2$, were proved impossible in Art. 7, it follows that $x^2+2y^2=nz^2$ and $x^2-2y^2=\pm nw^2$ must be also be impossible, whatever rational number n may be, and in like manner, $x^2+3y^2=nz^2$ and $x^2-3y^2=\pm nw^2$ must be impossible.

By taking $n=1$, and also $b=1$, we can form one solution of the equations $x^2+ay^2=z^2$ and $x^2-ay^2=w^2$ deduce another solution in greater integers; thus new $X=\frac{1}{2}(z^4+w^4)$ and new $Y=2xyzw$, as in case 2^0 Art. 9. *Ex, gr.* When $a=6$ then $x=5$ and $y=2$ give $z=7$ and $w=1$ $\therefore$ new $x=\frac{1}{2}(7^4+1^4)=1201$ and new $y=10\times 2\times 7=140$, giving $z=1249$ and $w=1151$, and thence again new $x=\frac{1}{2}(1249^4+1151^4)=$&c., and new $y=1201\times 280\times 1249\times 1151$ &c., &c. When $a=7$ then taking $n=2$, one obvious solution of the auxiliary equations $x^2+7y^2=2z^2$ and $x^2-7y^2=\pm 2w^2$ is $x=5$ $y=1$ $z=4$ and $w=3$ and $\therefore$ by this present Art. 10, new $x=4^4+3^4=337$ and new $y=2xyzw=120$ which are the least integral values of x and y to fulfil the proposed equations, $x^2+7y^2=z^2$ and $x^2-7y^2=w^2$ giving $z=463$ and $w=113$ and thence again we obtain, as directed above, new $x=\frac{1}{2}(463^4+113^4)$ and new $y=337\times 240\times 463\times 113$, from which we could again ascend to other values of x and y still larger to answer the proposed equations $x^2\pm 7y^2=\square^s$.

When $a=13$ then taking $n=1$ one obvious solution of the two *auxiliary* equations $x^2+13y^2=z^2$ and $x^2-13y^2=-w^2$ is $x=6$, and $y=5$ giving $z=19$ and $w=17$, and hence by this Art. 10, the values of x and y in the proposed equations, $x^2+13y^2=z^2$ and $x^2-13y^2=w^2$ are $x=\frac{1}{2}(19^4+17^4)=106921$, and $y=10\times 6\times 19\times 17=19380$, which are the least integral answers; these give $z=127729$ and $w=80929$, from which again we find new $x=\frac{1}{2}(127729^4+80929^4)$ and new $y=2\times 106921\times 19380\times 127729\times 80929$, &c., &c.

11.—The solution of the two equations, $X^2+abY^2=\square=Z^2$ and $X^2-abY^2=\square=W^2$ can be also derived from a solution of the *auxiliary* equations, $x^2+y^2=az^2$ and $x^2-y^2=bw^2$, for in fact $X=x^4+y^4$ and $Y=2xyzw$ will answer. Proof—For then $abY^2=4abx^2y^2z^2w^2=4x^2y^2(az^2)(bw^2)=4x^2y^2(x^4-y^4)=2tv$ where $t=x^4-y^4$ and $v=2x^2y^2$: and $X^2=(x^4+y^4)^2 \therefore =t^2+v^2$, and so $X^2\pm abY^2=(t\pm v)^2$ which are both squares, Q. E. D.

If a contains a square factor A^2, this factor may evidently be rejected, for by putting y' instead of Ay it is evident that $x^2\pm aA^2y^2=\square^s$ will be possible or impossible according as $x^2\pm ay'^2=\square^s$ are possible or impossible; now the two proposed equations $x^2+ay^2=z^2$ and $x^2-ay^2=w^2$ give $2x^2=z^2+w^2$ and $2ay^2=z^2-w^2$ and putting $z=z'+w'$ and $w=z'-w'$, these give $x^2=z'^2+w'^2$ and $ay^2=2z'w'$ the former is fulfilled, (Art. 1,) by $z'=m^2-n^2$ and $w'=2mn$ and then the latter can be fulfilled if $a=\dfrac{4mn\,(m+n)\,(m-n)}{y^2}$, y^2 being the greatest square factor contained in the numerator, and m as usual being prime to n; indeed it is thus Euler finds (in page 425 of his Algebra) the value of a so that the two proposed equations may be simultaneously possible; but $a=13$ would not be easily got by this method, $m=5$ and $n=4$ give $a=5$; $m=2$ and $n=1$ give $a=6$; $m=16$ and $n=9$ give $a=7$; $m=8$ and $n=1$ give $a=14$; $m=4$ and $n=1$ give $a=15$; $m=4$ and $n=3$ give $a=21$, $m=3$ and $n=2$ give $a=30$; $m=25$ and $n=9$ give $a=34$, &c, &c.

12.—The two simultaneous equations $\left\{\begin{matrix} x^2+10y^2=\square=z^2 \\ x^2-10y^2=\square=w^2 \end{matrix}\right\}$ are impossible.

By addition and subtraction $2x^2=z^2+w^2$ and $20y^2=z^2-w^2$ from the first 3 of these equations it follows by Art. 5, that if the proposed equations be possible they must be so when x, y, z, and w are all prime to each other, and, moreover, z and w and x prime to 10 and ∴ odd, so that we must as usual have $z=z'+w'$ and $w=z'-w'$, z' being prime to w', one even and the other odd; this substitution changes the 2 derived equations into $x^2=z'^2+w'^2$ and $5y^2=z'w'$ the general solution of the former (by Art. 1,) is $z'=m^2-n^2$, $w'=2mn$ and $x=m^2+n^2$, m being prime to n, one even and the other odd, and then the latter equation becomes $mn\,(m+n)\,(m-n)=\frac{5y^2}{2}$ which shows that y must be even, let $y=2y'$ and we get $mn\,(m+n)\,(m-n)=10y'^2$, which shows that some one of the 4 factors must be divisible by 10, (and of course it must be the even factor m or n,) or else one of them, (m or n,) must be divisible by 2, and another by 5, and then the quote or quotients being prime to each other, and to the remaining factors, and their continued product a square ∴ each quote and remaining factor must itself be a square. Now 1^0, I say that neither of the two odd factors $m\pm n$ can be divisible by 5; for if $m+n=5r^2$ and $m-n=s^2$, $m=p^2$ and $n=2q^2$ then $5r^2+s^2=2p^2$ an equation already proved impossible in case 3^0 Art. 9; and if it were $m+n=5r^2$ and $m-n=s^2$, $m=2p^2$ and $n=q^2$, then $5r^2-s^2=2q^2$, which is impossible for much the same reason as before; thus then $m+n$ cannot be divisible by 5; and if $m-n$ were so divisible we should have $m+n=r^2$, $m-n=5s^2$, and so we should have $5s^2=2p^2-q^2$ or else$=p^2-2q^2$ which are both impossible for reasons already assigned in case 3^0 Art. 9; and thus we must have $m+n=r^2$ and $m-n=s^2$, and this being so I say 2^0—that m and n cannot be one divisible by 5, and the other divisible by 2, for if so then we should have $r^2=2p^2+5q^2$ or else $=5p^2+2q^2$ which are both impossible for reasons assigned in preceding case, and ∴ if it be possible at all to have $mn\,(m+n)(m-n)=10y'^2$ under the conditions already mentioned, either m or n must be divisible by 10.

Now 3^0.—I say m cannot be ÷10, for if so we should have $m=10p^2$, $n=q^2$, $m+n=r^2$ and $m-n=s^2$ and ∴ $r^2+s^2=20p^2$ which is impossible as r and s must be both odd, and ∴ r^2+s^2 of the form $8A+2$ and $20p^2$ is not of this form. —4^0. Nor can n be ÷10 for if so we should have $p^2+10q^2=r^2$ and $p^2-10q^2=s^2$, a pair of equations precisely similar to the original pair, and as p and q are $<x$ and y. ∴ if the proposed equations admitted of a solution in great large integers, they should also admit of a solution in smaller integers, and thence in still smaller integers, &c., &c; and as there exists no solution in small numbers there can ∴ exist no solution among even the largest integers; the particular case of $q=0$ in $p^2\pm10q^2=\square$ would not enable us to ascend into large integers, for x and y in $x^2\pm10y^2=\square$ since $y=2pqrs$ would$=0$ if $q=0$, and so this particular case leads to no real solution, neither could we descend to it from any real solution, Q.E.D.

Cor.—Hence, also, the formulæ $nz^2=2x^2+5y^2$ and $\pm nw^2=\pm2x^2-5y^2$ must be impossible whatever rational n^o n may be, since by Art. 10, if these were possible the proposed formulæ $x^2 \pm 10y^2 = \square^s$ would be so too; and in like manner it follows from Art. 11 that the two simultaneous equations $x^2+y^2=2z^2$ and $x^2-y^2=5w^2$ are impossible, as are also the pair $x^2+y^2=5z^2$ and $x^2-y^2=2w^2$, and it is obvious that we could derive numerous other impossible pairs from the exceptional values of a given in Cor. 2, Art. 8; but as the pair $x^2 \pm ay^2 = \square^s$ are possible when $a=5$, 6, 7, 14, 15, or 21, &c.; ∴ by what is said in Art. 11 they will be also possible when $a=5^m \times 3^n$, or $=7^m \times 2^n$, or $7^m \times 3^n$ if m be odd when n is so.

Cor. 2—We may here remark that the proved impossibility of any pair of of simultaneous equations, as

$$\begin{Bmatrix} x^2-y^2=z^2 \\ x^2-3y^2=w^2 \end{Bmatrix}$$

involves with it the impossibility of five other pairs, which are easily found from the given pair; *Ex. gr.* solving the two equations given here, for x and z; x and w; y and z; &c., we find the five following pairs, each of which is impossible, since the above original pair known (and proved in Art. 32) to be impossible, may be re-derived from any of these five pairs, viz:—

$$\begin{array}{l|l|l|l|l} w^2+2y^2=z^2 & z^2+y^2=x^2 & x^2-w^2=3y^2 & z^2-w^2=2y^2 & x^2-z^2=y^2 \\ w^2+3y^2=x^2 & z^2-2y^2=w^2 & 2x^2+w^2=3z^2 & 3z^2-w^2=2x^2 & 3z^2-2x^2=w^2 \end{array}$$

and thus it follows, from Art 6, that the pairs

$$\left|\begin{array}{l|l} w^2+y^2=x^2 & z^2-y^2=x^2 \\ w^2+2y^2=z^2 & z^2-2y^2=w^2 \end{array}\right| \quad \text{are impossible.}$$

13—The single equation $ax^2+by^2=cz^2$, is always possible, if ac or bc be a square number. For by multiplying the equation by c and then suppressing the square factors as directed in Art. 4, the equation will evidently be brought to the general form $x^2-y^2=Az^2$ already solved in Art. 2, or as follows:—suppose A is fractional, and $=\frac{a}{b}$ then the equation gives A or $\frac{a}{b}=\frac{x^2-y^2}{z^2}=\frac{x+y}{z} \times \frac{x-y}{z}$ and ∴ if $\frac{x+y}{z}=\frac{p}{q}$ then $\frac{x-y}{z}$ must $=\frac{a\,q}{b\,p}$ and ∴ by adding and subtracting $\frac{x}{z}=\frac{b\,p^2+a\,q^2}{2bpq}$ and $\frac{y}{z}=\frac{b\,p^2-a\,q^2}{2bpq}$ so that $x=b\,p^2+a\,q^2, y=b\,p^2-a\,q^2$ and $z=2bpq$ will answer the proposed equation $x^2-y^2=Az^2=\frac{a}{b}z^2$ whatever rational numbers p and q may be,

14—But when ab is a square number, c'^2 the proposed equation $ax^2+by^2=cz^2$ will be possible or impossible, according as A is or is not the sum of two integral squares, where $A=\frac{a\,c}{p^2}$; p^2 being the *greatest* square factor contained in $a\,c$; for by multiplying the proposed equation, by a and suppressing the square factors as in Art. 4, by putting $x'=ax$ and $y'=c'y$ and $z'=p\,z$ we find $x'^2+y'^2=Az'^2$, in which the coeft. A contains no square factor: and now if A be the sum of two integral squares *i.e.* $A=p^2+q^2$ 'tis evident the derived equation $x'^2+y'^2=Az'^2$ is satisfied by taking (by Art. 1) $z'^2=r^2+s^2, x'=pr \pm qs$ and $y'=ps \mp qr$ from which the solution of the proposed equation can be obtained as in Art. 4.

But if $A=3$, 6, 7, 11, 12, 14, or any other integer which is not the sum of two integral squares, or when any such integer is a divisor of A, the derived equation $x^2+y^2=Az^2$ (and ∴ also the proposed equation) will be impossible

1°—$x^2+y^2=3z^2$ is impossible; for if it were possible it should be so, by Art. 5, when x, y, z, are all prime to each other; now x being prime to y, they cannot both be divisible by 3 and $\therefore x^2+y^2\equiv 3N+1$ or $3N+2$ according as one or none of x, y, is $\div 3$, but neither $3N+1$ or $3N+2$ when divided by 3 could give the integral quotient z^2.

2°—$x^2+y^2=6z^2$ is also impossible, since by 1° it is proved that x^2+y^2 cannot be divisible by 3 when x is prime to y.

3°—$x^2+y^2=7z^2$ is impossible; for every square number $\equiv 7N$, $7N+1$, $7N+2$, or $7N+4$, and $\therefore$ the sum of two squares (x^2+y^2) cannot be divisible by 7, when they are prime to each other and $\therefore$ not each divisible by 7—

4° $x^2+y^2=11z^2$ is also impossible; for every square number $=11N$, $11N+1$, $11N+3$, $11N+4$, $11N+5$, or $11N+9$, and $\therefore$ the sum of two square numbers (x^2+y^2) cannot be $\div 11$ when they are prime to each other, and $\therefore$ not each divisible by 11.

5°—$x^2+y^2=14z^2$ is also impossible, since by 3°, x^2+y^2 is not $\div 7$ when x is prime to y.

6°—$x^2+y^2=15x^2$ is also impossible, since by 1°, x^2+y^2 is not $\div 3$ when x is prime to y. In fact, if the equation $x^2+y^2=Az^2$ be possible, it must by Art. 5 be so, when x, y, z are all prime to each other, as A contains no square factor; then A (or z) will divide x^2+y^2 or $(x^2+y^2)(a^2+b^2)=(ax+by)^2+(bx-ay)^2$, whatever integers a and b may be; but as x is prime to y, a and b can be found such that $bx-ay=\pm 1$ and thus A (or z), will divide $(ax+by)^2+1$; now let q be the nearest integer to $\frac{ax+by}{A}$ whether greater or less, then $ax+by$ will $=Aq\pm p$ where p will not exceed $\frac{1}{2}A$; and as A must divide $(ax+by)^2+1=(Aq\pm p)^2+1=A^2q^2\pm 2Apq+p^2+1$ it must also divide p^2+1 and hence unless A divides p^2+1 and p not $>\frac{1}{2}A$, the equation $x^2+y^2=Az^2$ will be impossible. That this condition of p^2+1 being divisible by A is tantamount to saying that A must be the sum of two integral squares is thus proved by M. Hermite.

Convert $\frac{p}{A}$ into a continued fraction and then find the fractions converging to $\frac{p}{A}$, let $\frac{m}{n}$ and $\frac{m'}{n'}$ be two of these converging fractions, consecutive and such that n is less, and n' greater than $\sqrt{A}$. which is always possible since 1 and A are the denominators of the first and last converging fractions; the difference of $\frac{p}{A}$ and $\frac{m}{n}$ is, by the property of the converging fractions $<\frac{1}{nn'}$ and $\therefore (\frac{p}{A}-\frac{m}{n})^2<\frac{1}{n^2n'^2}$, or $(pn-Am)^2<\frac{A^2}{n'^2}$ but by hypothesis $n'^2>A$ $\therefore$ $(pn-Am)^2<A$ and as by hypothesis $n^2<A$ $\therefore$ $(pn-Am)^2+n^2<2A$ *i. e.* $n^2(p^2+1)-2Amnp+A^2m^2<2A$ and as the lett hand member is divisible by A for by hypothesis p^2+1 is divisible by A $\therefore$ we must have $(np-Am)^2+n^2=A$ and hence follows this Theorem, viz., "Every divisor A of the sum of two squares (x^2+y^2) that are prime to each other is itself sum of two squares."—

Cor. 1.—A being any whole number the equation $x^2+y^2=A$ will be impossible in fractions if it be so in integers; for if $x=\frac{p}{q}$ and $y=\frac{r}{s}$ the given equation becomes $p^2s^2+q^2r^2=Aq^2s^2$ which has just been proved impossible when A is not a square number, nor the sum of two integral squares.

Cor. 2.—When A is the sum of two integral squares prime to each other, then if r' be a possible remainder of squares to modulus or divisor A, I say A—r' will be also a possible remainder. For let $A=p^2+q^2$ then A divides

$(p^2+q^2)(r^2+s^2)=(pr+qs)^2+(ps-qr)^2$, but as p is prime to q.∴ r and s can be found such that $ps-qr$ can become = any given integer, and ∴ the remainder of $(ps-qr)^2 \div A$ can become any one of the possible remainders, but it is evident that if remainder of $(ps-qr)^2 \div A$ be r' the remainder of $(pr+qs)^2 \div A$ must be $A-r'$.∴ &c. Q.E.D.

Cor. 3.—Towards the end of the art. 10 we had to solve $x^2+13y^2=z^2$ and $x^2-13y^2=-w^2$; this latter giving $x^2+w^2=13y^2$ shows that y is a divisor of x^2+w^2 and must ∴ by this Art. 14 be itself the sum of two integral squares we ∴ try $y=1^2+2^2=5$, and then the first equation $13y^2=z^2-x^2$ gives $13\times 25=(z+x)(z-x)$ now the derived equation $2x^2=z^2-w^2$, shows, by Art. 5 that z and w must be prime to 2 and ∴ both odd and ∴ x even so that $z+x$ and $z-x$ are both odd and prime to each other since x, y, z, and w are, by Art. 5, all prime to each other; but it is evident that 13×25 cannot be resolved into any other factors prime to each other, ∴ we must have $z+x=25$ and $z-x=13$, and ∴ $z=19$, $x=6$, &c., &c.

15.—But when in the equation $ax^2+by^2=cz^2$ neither ab, ac, nor bc is a square number, we may then use the following general method of ascertaining its impossibility given in Articles 52 and 53 of Barlow's Theory of Numbers:—

Let a prime number A be modulus or divisor of square numbers, and let all the possible remainders of squares to this modulus be r_1, r_2, r_3, &c.; then since N^2 and $(A-N)^2$, or $(mA\pm N)^2$ when divided by A leave the same remainder, ∴ the number of *different* remainders can never exceed $\frac{1}{2}A$; let s_1, s_2, s_3, &c. be the remaining integers $<$ A and different from r_1, r_2, r_3, &c., then $Aq+r$ will denote a possible form of squares to modulus A and $Aq'+s$ will be an impossible form, and it is easily proved, (Barlow, Art. 52), that the product of a possible and impossible form to the same modulus *always* produces an impossible form; *Ex, gr* the possible forms of squares to modulus 5 are $5q\pm 1$, and the impossible forms are ∴ $5q'\pm 2$, and their product has evidently the impossible form. Again to modulus 7, the possible forms are $7q+1$, $7q+2$, and $7q+4$, and ∴ the impossible forms are $7q'+3$, $7q'+5$, and $7q'+6$, and the product of one of each has the impossible form. To show the use of this theorem take any equation as $5x^2+y^2=3z^2$. Now by Art. 5. If this equation be possible it must be so when y and z are prime to 5, then ℘ above $3z^2$ has the impossible form to modulus 5; and as y^2 has necessarily the possible form, ∴ $3z^2-y^2$ could not be divisible by 5 so as to give the integral quotient x^2 and ∴ the equation $5x^2+y^2=3z^2$ is impossible. Let us now take $7x^2+11y^2=13z^2$ which gives $x^2=\frac{13z^2-11y^2}{7}=z^2-y^2+\frac{6z^2-4y^2}{7}$ and as 6 is an impossible remainder of □ˢ to modulus 7, and 4 is a possible remainder; and moreover, if the equation be possible at all it should be so when y and z are prime to 7, and then ℘ above $6z^2$ has the impossible form, and $4y^2$ has, of course, the possible form, and ∴ $6z^2-4y^2$ could not be divisible by 7 and ∴ the proposed equation $7x^2+11y^2=13z^2$ is impossible.

Let us next take $7x^2+11y^2=23z^2$ which gives $x^2=3z^2-y^2+\frac{2z^2-4y^2}{7}$ but as here the two coefts. 2 and 4 in the numerator are both possible remainders of □ˢ to modulus 7 and ∴ the remainders of $2z^2\div 7$ and $4y^2\div 7$ will both be possible, ∴ we cannot in this way infer the impossibility of the proposed equation.

But by solving it for y^2 we get $y^2=2z^2+\frac{z^2-7x^2}{11}$ and here as the coeft. 7 is an impossible remainder of squares to modulus 11 to which x and z must be prime

∴ the remainder of $7x^2 \div 11$ is an impossible remainder of squares to modulus 11, and so it could not be the same as remainder of $z^2 \div 11$ which is necessarily possible, and thus it appears that the proposed equation is impossible; and we may observe that the impossibility of $x^2 = \frac{cz^2 - by^2}{a}$ can only be inferred with certainty when the remainders of $\frac{c}{a}$ and $\frac{b}{a}$ are the *one possible*, and the *other impossible* remainders of squares to modulus a; this example shows also that we should solve the equation for each of the 3 quantities x^2, y^2, z^2, and apply the test to each solution, as in the preceding case.

Let us now take $14x^2 + 6y^2 = 17z^2$; here as the coefts. 14 and 6 have a common factor 2, z must be even, put it ∴ $= 2z'$ and we get $7x^2 + 3y^2 = 34z'^2$, first this gives $x^2 = 4z'^2 + \frac{6z'^2 - 3y^2}{7}$ from which we cannot infer the impossibility as 6 and 3 are both impossible remainders of squares to modulus 7; solving for y^2 we get $y^2 = 11z'^2 - 2x^2 + \frac{z'^2 - x^2}{3}$ from which we cannot infer the impossibility; lastly, solving for z^2 we get $2z'^2 = \frac{7x^2 + 3y^2}{17}$ ∴ $2z'^2 - x^2 = \frac{3y^2 - 10z^2}{17}$ and as 3 and 10 are both impossible remainders of squares to modulus 17 ∴ we cannot infer the impossibility, thus we are led to suspect that $7x^2 + 3y^2 = 34z'^2$ may be possible; and if it be it must by Art. 5 be so when x, y, z', are all prime to each other, and moreover, x and y prime to 34, and ∴ both odd, and also z' prime to 3 and 7; in fact $x = 1$, $y = 3$ and $z' = 1$ obviously fulfil it since $7.\ 1^2 + 3.\ 3^2 = 34.\ 1^2$. Now multiply this last equation by the proposed, and remembering the formula $(ax^2 + by^2)(am^2 + bn^2) = (amx \pm bny)^2 + ab(nx \mp my)^2$ we get $(7x + 9y)^2 + 21(3x - y)^2 = (34z')^2$ which is of the form $X^2 - Y^2 = 21Z^2$ already considered in Article 3.

When we have two simultaneous equations like those in Art. 5 we must apply the tests in this and the preceding Art. to each of the 4 equations mentioned in Art. 5, and if our tests prove any one of the 4 equations to be impossible when solved relatively to x^2, y^2, or z^2 then the two equations proposed are *certainly impossible* or incompatible *jointly* though not perhaps *seperately*, and it will fully appear in the sequel that the two proposed equations may be simultaneously impossible even when the preceding tests fail in proving any one of the above mentioned 4 equations to be impossible.

16.—The two simultaneous equations $x^2 + ay^2 = \square$ and $x^2 - ay^2 = \square$ are impossible whenever a is a prime number of the form $4n + 3$, and $m^2 - 2$ not divisible by a, m being $< \frac{1}{2}a$. For instance, the proposed equations are impossible when $a = 11$, 19, or 43, but not when $a = 7$, 23, 31, or 47, since to these last moduli (7, 23, 31, &c.), 2 is one of the *possible* remainders of squares; it will be sufficient to prove this for the particular case of $a = 19$ since a good hint is enough. Ex uno disce omnes.

DEMONSTRATION.

If the pair of equations admitted of solution when $a = 19$ we should, as before, have y even, and $= 2y'$ and also $19y'^2 = mn(m + n)(m - n)$, m being prime to n, one even and the other odd, and ∴ the four factors m, n, $m \pm n$ all prime to each other and only one of them (m or n) even; now some one of the 4 factors must be divisible by 19 as their product is so divisible, and the quote and each of the 3 remaining factors must be squares, since they are prime to each other and their continued product a square (y'^2); 1° if m be $\div$ 19 we should have $m=19p^2$, $n=q^2$, $m+n=r^2$ and $m-n=s^2$ and $\therefore$ $19p^2=q^2+s^2$ which is impossible by Art. 14, since 19 is not the sum of two integral squares, and indeed any number of the form $4n+3$ cannot be the sum of two squares, for as $4n+3$ is odd $\therefore$ one of the squares shonld be even and the other odd; but the even square$\equiv4m^2$ and the odd square$\equiv4m'+1$ and $\therefore$ the sum of both$\equiv$ $4m'+1$.

2°. If $m+n$ be $\div$ 19 we should have $m=p^2$, $n=q^2$, $m+n=19r^2$ and $\therefore$ $p^2+q^2=19r^2$ which is impossible for the reasons just assigned in preceding case.

3°. If $m-n$ be $\div$ 19 we should have $m=p^2$, $n=q^2$, $m+n=r^2$ and $m-n$ $=19s^2$ giving $19s^2=r^2-2q^2$ or $s^2=\dfrac{r^2-2q^2}{19}$ which is impossible by Art. 15 since by hypothesis the coeft. 2 (of $2q^2$) is an *impossible* remainder of squares when divided by $19=a$.

4°. If n be $\div$ 19 we should have $p^2+19q^2=r^2$ and $p^2-19q^2=s^2$ two equations precisely similar to the original pair, and thereby indicating that if the proposed equations admitted of a solution in large integers there should also exist a solution in smaller integers, and thence again in integers still smaller, &c., &c.

This proof essentially depends on the coeft., 2 being an impossible remainder of squares to divisor 19 $(=a)$ and $\therefore$ m^2-2 not divisible by a (m of course being $<\frac{1}{2}a$) is an *essential* condition; so also it is *essential* that a be a prime number and not the sum of two squares; we may obviously unite this last condition to the other by stating that a must be a prime number, and such that neither m^2+1 nor m^2-2 must be divisible by a, m being $<\frac{1}{2}a$, and when this is so a cannot be the common difference of three square numbers in arithmetical progression, for x^2+ay^2, x^2 and x^2-ay^2 are still in arithmetical progression even when all are divided by y^2, &c., &c.

Chapter II.

On the possible and impossible cases of the two simultaneous equations,
$x^2+y^2=\square$ *and* $x^2+ay^2=\square$.

17. The pair of simultaneous equations $\begin{cases} x^2+y^2=\square=w^2 \\ x^2+3y^2=\square=z^2 \end{cases}$ are impossible.

By elimination of x or y we get $2y^2=z^2-w^2$ and $2x^2=3w^2-z^2$ and $\therefore$ by Art. 5, if the proposed equations were possible they should be so when x, y, z, and w are all prime to each other, and also z and w odd and $\therefore$ y even, as appears from $2y^2=z^2-w^2$; hence, by Art. 1, the general solution of the first equation is $x=m^2-n^2$, $y=2mn$, m being prime to n, one even and the other odd, and as y was proved even $\therefore$ by Art. 3, the general solution of the second equation $3y^2=z^2-x^2$ is $y=2pq$ and $\pm x=3p^2-q^2$, p being prime to q, one even and the other odd; and equating the values of y we must have $mn=pq$, now this equation can subsist only thus, $m=ab$, $p=ac$, $q=bd$ and $\therefore$ $n=cd$ where a denotes the greatest C M of m and p, &c., and here a, b, c, d, must be all prime to each other (else m would not be prime to n, and p to q), and one of a, b, c, d, even else m, n, p, and q would be all odd; then $m^2-n^2=x=+(3p^2-q^2)$ gives $a^2b^2-c^2d^2=3a^2c^2-b^2d^2$, or $\therefore \dfrac{a^2+d^2}{3a^2+d^2}=\dfrac{c^2}{b^2}$ and as in the fraction $\dfrac{d^2+a^2}{d^2+3a^2}$ denominator—numerator$=2a^2$ and 3 numerator—denominator$=2d^2$ and as a is prime to d $\therefore$ $2a^2$ and $2d^2$ can have no C M but 2 $\therefore$ the fraction $\dfrac{d^2+a^2}{d^2+3a^2}$ is either in its lowest terms already, or only admits 2 as C M; but 2 cannot be C M, for if it were we should have $d^2+3a^2=2b^2$ which was already proved impossible in Art. 8, case 4°; and if the fraction were irreducible we should have $d^2+a^2=c^2$ and $d^2+3a^2=b^2$ which are exactly similar to the proposed equations, but having d and a instead of x and y, and as $x=m^2-n^2=a^2b^2-c^2d^2=(ab+cd)(ab-cd)$ is $>d$ and $y=2mn=2abcd$ is $>a$ $\therefore$ if the proposed equations admitted of a solution in great large integers, they should also admit of a solution in smaller, and thence in still smaller integers, &c., &c.; and as there exists no solution in small integers $\therefore$ there can exist no solution whatever in rational numbers. Q.E.D.

If we had taken $x=-(3p^2-q^2)$ then $3p^2-q^2=-x \therefore =n^2-m^2$ gives $\dfrac{b^2+c^2}{b^2+3c^2}=\dfrac{a^2}{d^2}$ which is exactly similar to the equation already obtained and $\therefore$ impossible for similar reasons. See *Euler's Algebra*, Art. 230.

18.—The pair of simultaneous equations $\begin{cases} x^2+y^2=\square \\ x^2+4y^2=\square \end{cases}$ are impossible. If they were possible they should obviously be so when x is prime to y, and as x^2+y^2 could not be a square if x and y were both odd, $\therefore$ one and only one of them must be even, and we may suppose it is y that is so, for if x were even and $=2x'$ our two equations would become $y^2+x'^2=\square$ and $y^2+4x'^2=\square$ and y being odd this $\therefore$ returns to the case or supposition of x odd in the two primitive equations and this case is $\therefore$ the only one that requires proof; now by Art. 1 the general solution of the first equation is $x=p^2-q^2$ and $y=2pq$, p being prime to q, one even and the other odd and as x is odd and prime to y

$\therefore$ it is prime to $2y$ and $\therefore$ by Art. 1 the general solution of $x^2+(2y)^2=\square$ is $x=m^2-n^2$ and $2y=2mn$, m being prime to n one even and the other odd $\therefore$ $mn=2pq$; here if m be odd this equation shows that its prime factors must be found among those of p and q $\therefore$ the equation can subsist only thus, $m=ab$, $p=ac$, $q=bd$, and $\therefore$ $n=2cd$, and then $m^2-n^2=x=p^2-q^2$ gives $a^2b^2-4c^2d^2=a^2c^2-b^2d^2$ or $\therefore$ $\frac{a^2+d^2}{a^2+4d^2}=\frac{c^2}{b^2}$ a, b, c, d, being all prime to each other and one of them even and $\therefore$ $\frac{c^2}{b^2}$ is only the fraction $\frac{a^2+d^2}{a^2+4d^2}$ reduced to its lowest terms; now denominator—numerator$=3d^2$ and 4 numerator—denominator $=3a^2$ and as a is prime d $\therefore$ the fraction $\frac{a^2+d^2}{a^2+4d^2}$ could have no C M but 3, but even 3 cannot be a C M since a^2+d^2 is never divisible by 3 when a is prime to d (Art 8, Case 1°) thus the fraction $\frac{a^2+d^2}{a^2+4d^2}$ is already in its lowest terms and $\therefore$ we must necessarily have $a^2+d^2=c^2$ and $a^2+4d^2=b^2$, a pair of equations precisely similar to the primitive pair $x^2+y^2=\square$ and $x^2+4y^2=\square$, but having a and d instead of x and y; now $y=2pq=2abcd$ is $>d$ and $x=p^2-q^2=a^2c^2-b^2d^2=(ac+bd)(ac-bd)$ is $>a$ so that if the two equations proposed could have a solution in large integers for x and y they should also be fulfilled by smaller integers (a and d) for x and y, and thence again by still smaller integers for x and y, &c., &c., and as there exists no solution in small numbers $\therefore$ there can be none amongst even the largest numbers; if n (and not m) were odd in the equation $mn=2pq$ then the 2nd equation $m^2-n^2=p^2-q^2$ could be written $n^2-m^2=q^2-p^2$ so that the conditions and demonstration would be exactly the same as before. Q.E.D.

Cor.—Hence also $y^4-y^2+1=\square$ is impossible; for if it were possible we should, by putting $x=y^2-1$, obviously have $x^2+y^2=\square$ and $x^2+4y^2=\square$ which have been just proved impossible in integers and $\therefore$ (by Art. 5) impossible also in fractions. And in like manner the impossibility of $y^4+y^2+1=\square$ follows from that of the two simultaneous equations, $x^2-y^2=\square$ and $x^2-4y^2=\square$ where $x=y^2+1$. See Art. 33.

Cor, 2.—From Articles 17 and 18 it follows by the method in Cor. 2, Art. 12, that each of the nine following pairs is impossible, viz.:—

$w^2-y^2=x^2$	$z^2-x^2=3y^2$	$z^2-2y^2=w^2$	$z^2-w^2=2y^2$	$w^2-x^2=y^2$
$w^2+2y^2=z^2$	$z^2+2x^2=3w^2$	$z^2-3y^2=x^2$	$3w^2-z^2=2x^2$	$3w^2-2x^2=z^2$
$w^2-y^2=x^2$	$z^2-3y^2=w^2$	$z^2-w^2=3y^2$	$w^2-x^2=y^2$	
$w^2+3y^2=z^2$	$z^2-4y^2=x^2$	$4w^2-z^2=3x^2$	$4w^2-3x^2=z^2$	

Cor. 3.—There cannot be four square numbers in arithmetical progression; if possible let w^2, x^2, y^2 and z^2, be four such square numbers, w^2 being the least and z^2 the greatest, we should then have $\begin{Bmatrix}2x^2-y^2=w^2\\2y^2-x^2=z^2\end{Bmatrix}$ take $x'=y^2z^2-x^2w^2=y^2(2y^2-x^2)-x^2(2x^2-y^2)=2y^4-2x^4$ and $y'=2xyzw$, and then it is evident that $x'^2+y'^2=(y^2z^2+x^2w^2)^2$; and by eliminating w and z from the value of $x'^2+4y'^2$ by means of the two original equations we find $x'^2+4y'^2=4(y^4-x^4)^2+16x^2y^2z^2w^2=4(y^4-x^4)^2+16x^2y^2(5x^2y^2-2x^4-2y^4)$ $\therefore$ $=4(y^4-4x^2y^2+x^4)^2$ and thus if there could be four square numbers in arithmetical progression the two equations $x'^2+y'^2=\square$ and $x'^2+4y'^2=\square$ would be simultaneously possible, but this cannot be by Art. 18 $\therefore$ &c.

19.—From observing that the equation $\frac{a^2+d^2}{a^2+3d^2}=\frac{c^2}{b^2}$ in Art. 17, would be fulfilled if we could have $a^2+d^2=nc^2$ and $a^2+3d^2=nb^2$, and that we would then have $x=a^2c^2-b^2d^2$ and $y=2abcd$, to fulfil the equations $\left\{\begin{matrix} x^2+y^2=w^2 \\ x^2+3y^2=z^2 \end{matrix}\right\}$ and from a perfectly similar remark on the final equation $\frac{a^2+d^2}{a^2+4d^2}=\frac{c^2}{b^2}$ in Art. 18, where we would then have $x=a^2c^2-b^2d^2$ and $y-2abcd$ to answer the equations $\left\{\begin{matrix} {}^2+y^2=w^2 \\ x^2+4y^2=z^2 \end{matrix}\right\}$ we are easily led to the invention of the following

GENERAL THEOREM—

The values of X and Y in $X^2+Y^2=\square=Z^2$, and $X^2+abY^2=\square=W^2$ can be deduced or obtained from the values of x and y in the *auxiliary* equations $x^2+ay^2=nz^2$ and $y^2+bx^2=nw^2$, in fact, I say, that $X=x^2w^2-y^2z^2$ and $Y=2xyzw$ will answer.

DEMONSTRATION—

For then 'tis evident that $X^2+Y^2=(x^2w^2+y^2z^2)^2$ and so the first condition is fulfilled; now $nX=x^2(y^2+bx^2)-y^2(x^2+ay^2)$ $\therefore$ $=bx^4-ay^4$; also $n^2Y^2=4x^2y^2(x^2+ay^2)(y^2+bx^2)$ $\therefore$ $=4x^4y^4(1+ab)+4bx^6y^2+4ax^2y^6$ and so $n^2.(X^2+abY^2)=b^2x^8-2abx^4y^4+a^2y^8+4ab(1+ab)x^4y^4+4ab^2x^6y^2+4a^2bx^2y^6=(bx^4+2abx^2y^2+ay^4)^2$, and hence $X^2+abY^2=(bx^4+2abx^2y^2+ay^4)^2\div n^2$ $\therefore$ $=\square$ and thus these values of x and y satisfy the second condition also Q E D.

By taking $a=b=-2$ and $n=-1$. This general theorem shows that the solution of $x^2+y^2=\square$ and $x^2+4y^2=\square$ could be obtained from the solution of $2y^2-x^2=z^2$ and $2x^2-y^2=w^2$, as was proved in the preceding Cor. 3; but without limiting the value of n (supposed negative) 'tis evident from this general theorem, that the solution of $x^2+y^2=z^2$ and $x^2+4y^2=w^2$ could be obtained from the solution of $2y^2-x^2=nz^2$ and $2x^2-y^2=nw^2$, and as the former pair of equations has been already proved to be impossible $\therefore$ the latter ($2y^2-x^2=nz^2$, and $2x^2-y^2=nw^2$) are also impossible *simultaneously*, whatever rational number n may be—

By taking $a=b=2$ and n positive, we see that $x^2+2y^2=nz^2$, and $y^2+2x^2=nw^2$ must be also impossible together, as otherwise, the pair $x^2+y^2=\square$ and $x^2+4y^2=\square$ would be possible, as proved in this general theorem; see Cor. 1, Art. 8.—In some of the subsequent articles it is proved that $x^2+y^2=\square$ and $x^2+Ay^2=\square$ are impossible *jointly* (not *separately*) when A is any whole number <20 except 7, 10, 11, or 17 whence we can infer by our general theorem the impossibility of transforming simultaneously into squares, an immense number of pairs of quadratic formulae of the form ax^2+by^2, and we have already observed Cor. 2, Art. 12, that the proved impossibility of any pair involves with it the impossibility of 5 other pairs which are easily found from the given pair *Ex, gr.* as the two equations $x^2+y^2=\square$ and $x^2+6y^2=\square$ are impossible (Art, 23) $\therefore$ by our general theorem, the pair $(x^2\pm2y^2)n=\square$ and $(y^2\pm3x^2)n=\square$ are also impossible *together* whatever rational number n may be; and by conceiving n negative when the lower sign (—) is used, it follows also from our general theorem that $2y^2-x^2=nz^2$ and $3x^2-y^2=nw^2$ are impossible n being *any* rational number.

Cor.—By taking $b=1$ and interchanging z and w in our general theorem, we see that the solution of $X^2+Y^2=Z^2$ and $X^2+aY^2=W^2$ can be obtained from the solution of $x^2+y^2=nz^2$ and $x^2+ay^2=nw^2$, merely by taking $X=y^2w^2-x^2z^2$ and $Y=2xyzw$; and then again, by taking $n=1$, this general theorem shows us how to find a solution in great *wh.* numbers from a known solution, in smaller integers, of $x^2+y^2=\square=z^2$ and $x^2+ay=\square=w^2$, for then new $X=y^2w^2-x^2z^2 \therefore =ay^4-x^4$ and new $Y=2xyzw$ in all cases; *Ex. gr.*—let $a=7$, so that the two equations to be solved are $x^2+y^2=\square=z^2$ and $x^2+7y^2=\square=w^2$ taking $n=2$, we see that one obvious solution of the two auxiliary equations $x^2+y^2=2z^2$ and $x^2+7y^2=2w^2$ is $x=y=z=1$ and $w=2$ and $\therefore$ ℘ above, $X=y^2w^2-x^2z^2=3$ and $Y=2xyzw=4$ is a solution of the two proposed equations, and now, using these values of X and Y for x and y we thence get another solution as indicated above, viz., new $X=ay^4-x^4=7.4^4-3^4=1711$, and new $Y=2xyzw=1320$ which may be now used for x and y, and thence a new solution in still larger integers can be again obtained, &c. &c.

As another example, let $a=17$, so that the two equations proposed, are $x^2+y^2=\square=z^2$ and $x^2+17y^2=\square=w^2$, and taking as above $n=2$ viz., a possible remainder of squares to modulus 17 or 7 (see Art. 15) we see that one obvious solution of the two auxiliary equations $x^2+y^2=2z^2$, and $x^2+17y^2=2w^2$ is $x=y=z=1$ and $w=3$ and $\therefore$ by the above Cor. $X=y^2w^2-x^2z^2=8$ and $Y=2xyzw=6$, or taking their halves $x=4$ and $y=3$ is a solution of the two proposed equations whence another solution is immediately got by the method indicated in the Cor. viz. new $X=ay^4-x^4=1121$ and new $Y=2xyzw=1560$, which numbers may be now used for x and y, and thence a new solution in still larger integers can be again obtained by the same method, &c. &c.

As a third example, let $a=11$ then taking $n=5$, viz., a *possible* remainder of squares to modulus 11, and also the sum of two squares (see Articles 15 and 14) we see that one obvious solution of the two auxiliary equations $x^2+y^2=5z^2$, and $x^2+11y^2=5w^2$ is $x=z=1$, $y=2$ and $w=3$, and so ℘ Cor. $X=y^2w^2-x^2z^2=35$ and $Y=2xyzw=12$ and these are the *least* values of x and y to fulfil the two proposed equations $x^2+y^2=z^2$ and $x^2+11y^2=w^2$, they give $z=37$ and $w=53$, and thence again by the Cor. new $X=11y^4-x^4 \therefore =1272529$ and new $Y=2xyzw=70\times12\times37\times53=1647240$ and thence again new $X=11y^4-x^4=11.1647240^4-1272529^4$, &c., &c.

As a last example let $a=10$ then the equations to be solved are $x^2+y^2=\square=z^2$ and $x^2+10y^2=\square=w^2$, here one obvious solution is $x=3$ and $y=4$ giving $z=5$ and $w=13$ and hence by the foregoing Cor. new $X=10y^4-x^4=2479$, and new $Y=2xyzw=1560$, &c., &c. And if we had to solve $x^2+y^2=z^2$ and $x^2+20y^2=w^2$, then taking $20=-2\times-10$ and $n=-1$ we have first to solve the two auxiliary equations, $x^2-2y^2=-z^2$ and $y^2-10x^2=-w^2$ whose solution is plainly $x=y=z=1$ and $w=3$, and hence, by our general theorem, new $X=x^2w^2-y^2z^2=8$ and new $Y=2xyzw=6$, or taking their halves $x=4$ and $y=3$ fulfil the proposed equations, giving $z=5$ and $w=14$ and thence again by the Cor. new $X=20.3^4-4^4=4\times341$, and new $Y=2xyzw=4\times420$, so that $x=341$ and $y=420$ will do, and thence again by the Cor. new $X=20.420^4-341^4$, &., &.

20. The two auxiliary equations $x^2+y^2=nz^2$ and $x^2+ay^2=nw^2$, by taking $y=z=1$, $x=B$ and $w=A$ give $n=B^2+1$ and $\therefore a=(B^2+1)\,A^2-B^2$ so that whenever $a=(B^2+1)\,A^2-B^2=A^2B^2\pm(A^2-B^2)$ then the proposed equations $x^2+y^2=\square=z^2$ and $x^2+ay^2=\square=w^2$ will be simultaneously possible; for then the auxiliary equations are obviously fulfilled by the above mentioned par-

ticular solution, viz.: $y=z=1$, $x=B$ and $w=A$, which by Cor., Art. 19, give new $X=y^2w^2-x^2z^2 \therefore =A^2-B^2$ and this value of x in the proposed equations would $=0$ if $B=A$, and thence every derived new $Y'=2XYZW$ would$=0$ which would lead to no real solution (nor useful result) of the proposed equations; hence we must be careful not to take $B=A$ (nor $\therefore a=A^4$) in $a=A^2B^2 \pm(A^2-B^2)$; $B=1$ gives $a=2A^2-1$, which includes the integers 7, 17, 31, 49, 71, 97, 127, &c.; $B=2$ gives $a=5A^2-4$ or $3A^2+4$ which includes the integers 41, 52, 76, 79, 112, 121, &c.; $B=3$ gives $a=10A^2-9$ or $8A^2+9$. In fact we see that—

2 & 3 for A & B (or for B & A) give	$a=A^2B^2\pm(A^2-B^2)$	$=36\ \pm5$	$=31$	or 41
2 & 4	$a=$ ...	$=64\ \pm12$	$=52$	or 76
2 & 5	$a=$ ...	$=100\pm21$	$=79$	or 121
2 & 6	$a=$ ...	$=144\pm32$	$=112$	or 176
3 & 4	$a=$ ...	$=144\pm7$	$=137$	or 151
				&., &c.

The same auxiliary equations, by taking $y=2$, $z=1$, $x=B$ and $w=A$ give $n=B^2+4$ and $4a=(B^2+4)A^2-B^2$ and here if B were even and $=2B'$ we would get $a=(B'^2+1)A^2-B'^2$ which brings us back to the preceding case already considered, we $\therefore$ take B odd, and 1°, $B=1$ gives $a=\dfrac{5A^2-1}{4}$ which includes the integers 11, 31, 61, 101, 151, &c., &c.

2°. $B=3$ gives $a=\dfrac{13A^2-9}{4}$ which includes the integers 27, 79, 157, &c. The same auxiliary equations by taking $y=3$, $z=1$, $x=B$, $w=A$ give $n=B^2+9$ and $a=A^2+B^2\times\dfrac{A^2-1}{9}$ and then $A=8$ gives $a=7B^2+64$ which includes the integers 71, 92, &c., and $A=10$ gives $a=11B^2+100$ which contains the integers 111, 144, 199, &c.

Again, taking $y=4$ and $z=1$, $x=B$ which must be odd in order to be prime to y, and $w=A$, the same *auxiliary* equations give $n=B^2+16$ and $a=A^2+B^2\times\dfrac{A^2-1}{16}$ and here $A=7$ gives us $a=3B^2+49$ and $A=9$ gives $a=5B^2+81$, which contains 86, 124, 126, &c., besides some of the integers already otherwise found for a. . . . By taking $y=6$ and $z=1$ we would find in a similar manner that a could be $=8B^2+289$ or $=10B^2+361$.

The *auxiliary* equations are evidently satisfied by the particular solution $x=4$, $y=3$, $z=5$ and $n=1$ if $a=\dfrac{w^2-16}{9}$; and to have a integral take $w=9n\pm4$ and $\therefore a=9n^2\pm8n$ which includes the integers 17, 20, 52, 57, 105, 112, 176, &c., some of which values of a have been already found otherwise. The auxiliary equations are also evidently fulfilled by the particular solution $x=3$, $y=4$, $z=5$, $n=1$, if $a=\dfrac{w^2-9}{16}$ and to have a integral take $w=8n\pm3$ and $\therefore a=4n^2\pm3n=7$, 10, 22, 27, 45, 52, 76, 85, 115, 126, 152, &c. In like manner $x=12$, $y=5$, $z=13$, $n=1$ and $w=25n\pm12$ would give $a=25n^2\pm24n=49$, 52, 148, 153, &c. The *same* auxiliary equations are obviously satisfied by $x=5$, $y=12$, $n=1$, if $a=\dfrac{w^2-25}{144}$ and $\therefore$ if $w=72n\pm5$ then $a=36n^2\pm5n=31$, 41, 134, 154, &c.; but if we take $w=72n\pm13$

then $a=36n^2\pm13n+1=24$, 50, 119, 171, &c. The auxiliary equations are also satisfied by $x=15$, $y=8$, $n=1$, if $a=\frac{w^2-225}{64}$ and $\therefore$ if $w=32n\pm15$ we get $a=16n^2\pm15n=31$, 34, 94, 99, 189, 196, &c.

21.—In the auxiliary equations, $x^2+ay^2=nz^2$ and $y^2+bx^2=nw^2$ of the general Theorem in Art. 19, let $x=y=z=1$ $\therefore$ by 1st Equation $n=a+1$ which substituted in the 2nd equation gives $1+b=(1+a)w^2$ and $\therefore$ $ab=a(a+1)w^2-a$ so that whenever $a'=a(a+1)A^2-a$ then the two equations $x^2+y^2=\square$ and $x^2+a'y^2=\square$ will be simultaneously possible; we must not, however, take $A=1$ (nor $\therefore$ $a'=a^2$) for as x, y, z, are taken each $=1$, if w (or A) were also $=1$ these values would give new $X(=x^2w^2-y^2z^2)=0$ from which no new solution could be derived: if $a=1$, then $a'=2A^2-1$ a formula already obtained in the preceding Art. 20; if $a=-2$ then $a'=2A^2+2$, but if $a=2$ then $a'=6A^2-2$ which contains the integers 22, 52, 94, 148, &c.; if $a=3$ then $a'=12A^2-3$ and $A>1$ gives $a'=45$, 105, 189, &c.; $a=4$ gives $a'=76$, 176.

But if in these two auxiliary equations we take $z=w$ and then eliminate n, we get $(a-1)y^2=(b-1)x^2$ and $\therefore$ if $a=1+mx^2$ then must $b=1+my^2$ and thence $a'=ab=(1+mx^2)\,(1+my^2)$ but as z was supposed $=w$, $\therefore$ x must not be taken $=y$ in the value of a', else new $X=y^2w^2-x^2z^2$ would $=0$ from which no new solution could be found, nor useful result derived; so that whenever $a'=(1+mA^2)\,(1+mB^2)$ and A not $=$B and $\therefore$ a' not a square number then the two proposed equations $x^2+y^2=\square=z^2$ and $x^2+a'y^2=\square=w^2$ will be simultaneously possible; if $m=1$ then $a'=(1+A^2)(1+B^2)$ and here $B=1$ and A not $=1$ gives $a'=2(A^2+1)$ which contains the integers 10, 20, 34, 52, 74, 100, &c., &c.; and $B=2$ gives $a'=5\,(A^2+1)$ and here A not $=2$ gives $a'=10$, 50, 85, 130, &c.; again $m=-1$ gives $a'=(A^2-1)\,(B^2-1)$ and here $B=2$ gives $a'=3(A^2-1)$ which contains the integers, 24, 45, 72, 105, 144, &c.; again $m=-2$ gives $a'=(2A^2-1)\,(2B^2-1)$ and here $B=1$ gives $a'=2A^2-1$ a formula already obtained otherwise; but $B=2$ gives $a'=7(2A^2-1)=7$, 119 217, &c., &c.; Again, $m=2$ gives $a'=(2A^2+1)(2B^2+1)$ and then $B=1$ gives $a'=3\,(2A^2+1)$ and $A>1$ gives $a'=27$, 57, 99, 153, &c. If only $z=1$ the first auxiliary Equation gives $n=x^2+ay^2$ and thence the second equation gives $b=w^2+\frac{y^2}{x^2}(aw^2-1)$ and if we take $a=2$, $w=5$, $x=7$, this gives $b=25+y^2$ and $\therefore$ $a'=ab$ $\therefore=2y^2+50$ and then new $X=x^2w^2-y^2z^2=35^2-y^2$ and $\therefore$ whenever $a'=2A^2+50$ and A not $=35$ the two proposed equations $x^2+y^2=\square$ and $x^2+a'y^2=\square$ are possible *together*, thus a' may be 52, 58, 68, 82, 100, &c., &c.

By taking $x=1$ and $a=2$ our first auxiliary equation $x^2+ay^2=nz^2$ becomes $1+2y^2=nz^2$ and this is fulfilled by $y=2$, $z=3$, and $n=1$, and then the 2nd auxiliary equation $y^2+bx^2=nw^2$ becomes $4+b=w^2$ giving $b=w^2-4$ and $\therefore$ $ab=a'=2w^2-8$ where w must not be taken $=6$, else new X would $=0$; this formula $a'=2A^2-8$ contains the integers 10, 24, 42, 90, 120, &c.

Again, $x=1$ and $a=3$, then the 1st auxiliary equation becomes $1+3y^2=nz^2$ which is fulfilled by $y=4$, $z=7$, and $n=1$, and then the 2nd auxiliary equation, $y^2+bx^2=nw^2$ becomes $16+b=w^2$ and $\therefore$ $b=w^2-16$ and $\therefore$ $a'=ab$ $\therefore=3A^2-48$ and (A not $=28$) gives $a'=27$, 60, 99, 144, &c.

Again taking $x=2$ and $a=5$ the first auxiliary equation $x^2+ay^2=n^2$ is fulfilled by $y=1=n$, $z=3$, and then the 2nd auxiliary equation, $y^2+bx^2=nw^2$ gives

$1+4b=w^2$ $\therefore$ $b=\frac{w^2-1}{4}$ and then $a'=ab$ $\therefore$ $=\frac{5}{4}(w^2-1)=10,\ 30,\ 60,\ 100$, &c.; the 1st auxiliary equation is also fulfilled by $x=2$, $y=3$, $a=5$, $n=1$, $z=7$, and then the 2nd gives $b=\frac{w^2-9}{4}$ and $\therefore$ $a'=ab=\frac{5}{4}(w^2-1)-10$ $\therefore$ $=20$, 50, 90, &c. The 1st auxiliary equation is also satisfied by $a=5$, $n=1$, $z=9$, $y=4$, and $x=1$, and then the 2nd equation gives $b=w^2-16$ and $\therefore$ $a'=ab=5(w^2-16)=45,\ 100,\ 165$, &c. We found $b=w^2+\frac{y^2}{x^2}(aw^2-1)$ and $\therefore$ if $a=1$ $w=3$ and $x=2$, then $b=9+2y^2$ and $\therefore$ $a'=ab=2A^2+9=11,\ 17,\ 41,\ 59,\ 107,\ 137$, &c., but if $a=1$, $w=7$, and $x=4$ then $b=a'=3y^2+49$, a formula already found; also $a=1$, $w=9$ and $x=4$ gives $b=a'=5y^2+81$ another formula found already; again $x=1$, $y=7$, $z=5$, $a=1$ and $n=2$ fulfil the 1st auxiliary equation, and then the 2nd equation gives $b=a'=2w^2-49$, &c.

By collecting all the results of Articles, 20 and 21, it follows that the two equations $x^2+y^2=\square$ and $x^2+ay^2=\square$ will be simultaneously possible whenever $a=2A^2-49$, $2A^2-8$, or $2A^2-1$, or $2A^2+2$, or $2A^2+9$, or $2A^2+50$, or whenever $a=3A^2-48$, $3A^2-3$, or $3A^2+4$, or $3A^2+49$, or whenever $a=5A^2-80$ or $5A^2-4$, or $5A^2+5$, or $5A^2+81$, or $6A^2-2$, or $6A^2+3$, or $4A^2\pm3A$, or $9A^2\pm8A$, or whenever a is $=\frac{5A^2}{4}$ diminished either by $\frac{1}{4}$ or $1\frac{1}{4}$, or $11\frac{1}{4}$, or $31\frac{1}{4}$, or in fact whenever a is any of the following integers, 39 of which are <100, viz., 7, 10, 11, 17, 20, 22, 23, 24, 27, 30, 31, 34, 41, 42, 45, 49, 50, 52, 57, 58, 59, 60, 61, 68, 71, 72, 74, 76, 77, 79, 82, 85, 86, 90, 92, 93, 94, 97, 99, 100, 101, 104, 105, 111, 112, 113, 115, 119, 120, 121, 122, &c., &c.

22.—The two simultaneous equations, $\left\{\begin{matrix} x^2+y^2=\square=z^2 \\ x^2+5y^2=\square=w^2 \end{matrix}\right\}$ are impossible.

By Art. 5, if these two equations were possible they should be so when x is prime to y, and $\therefore$ all prime to each other except perhaps z and w; and as by Art. 1, z must be odd in the 1st equation, $\therefore$ the derived equation $4y^2=w^2-z^2$ shows that w must also be odd and y even (since every odd square $=8n+1$); hence by Art. 1 the general solution of the first equation is $x=m^2-n^2$ and $y=2mn$, m being prime to n, one even and the other odd; and by Art. 3 the general solution of the 2nd equation is $\pm x=5p^2-q^2$ and $y=2pq$; p being prime to q, one even and the other odd; hence equating the values of y we must have $mn=pq$, which can subsist only thus: $m=ab$, $p=ac$, $q=bd$, and $\therefore n=cd$, a, b, c, d being all prime to each other, ond one of them even, (see Art. 17), then $m^2-n^2=x=\pm(5p^2-q^2)$ gives, by taking the sign $+$, $a^2b^2-c^2d^2=5a^2c^2-b^2d^2$ or $\therefore$ $\frac{d^2+a^2}{d^2+5a^2}=\frac{c}{b^2}$ Now denominator — numerator of left hand number $=4a^2$, and 5 numerator — denominator $=4d^2$, and as a is prime to d, $4a^2$ and $4d^2$ can have no CM but 2 or 4; but 2 could not be CM, for if it were we should have $d^2+5a^2=2b^2$ which is impossible (by Art. 15 or Art. 9, case 3°), and 4 could not be CM since a is prime to d, and $\therefore$ a^2+d^2 is of the form $4n+1$ or $4n+2$ according as one or both of a and d are odd; lastly, if the fraction $\frac{d^2+a^2}{d^2+5a^2}$ be irreducible we should have $d^2+a^2=c^2$ and $d^2+5a^2=b^2$, which being exactly similar to the proposed pair of equations we prove by the usual argument that the thing is impossible in large integers as there exists no solution in small integers, see Articles 17 and 18. By taking the sign — then $m^2-n^2=-(5p^2-q^2)$ can be written $n^2-m^2=5p^2-q^2$ giving $\frac{b^2+c^2}{b^2+5c^2}=\frac{a^2}{d^2}$ so that the conditions and demnostration are the same as before. Q.E.D.

23.—The two simultaneous equations $\left\{\begin{matrix} x^2+y^2=\square=z^2 \\ x^2+6y^2=\square=w^2 \end{matrix}\right\}$ are impossible.

By eliminating x, y, $5y^2=w^2-z^2$ and $5x^2=6z^2-w^2$, and as no coeft. of these four equations contains a square factor ∴ by Art. 5, if the equations be possible at all they must be so when x, y, z, w, are all prime to each other, and also x and w prime to 6 (coeft. of y^2 in the 2nd equation), and ∴ both odd and ∴ by 2nd equation y even, and the other three odd, as they are prime to y; hence by Art. 1 the general solution of the 1st equation is $x=m^2-n^2$, $y=2mn$, and $z=m^2+n^2$; and by Art. 3 the general solution of the derived equation $5y^2=w^2-z^2$ is $y=2pq$, and $\pm z=5p^2-q^2$, m being prime to n, and p to q, one of each pair being even and the other odd; hence we must have $mn=pq$ which can subsist only thus: $m=ab$, $p=ac$, $q=bd$, and ∴ $n=cd$, $abcd$ being prime to each other, and one of them even, then $m^2+n^2=z=\pm(5p^2-q^2)$ gives, by taking the sign $+$, $a^2b^2+c^2d^2=5a^2c^2-b^2d^2$, or ∴ $\frac{c^2+b^2}{5c^2-b^2}=\frac{a^2}{d^2}$ Now numerator $+$ denominator of left hand member $=6c^2$, and 5 numerator — denominator $=6b^2$, and ∴ as b is prime to c, $\frac{c^2+b^2}{5c^2-b^2}$ can have no CM except 6 or a divisor of 6; if 2 were common measure we should have $c^2+b^2=2a^2$ and $5c^2-b^2=2d^2$ giving $3c^2=a^2+d^2$ which is impossible by Art. 14, 1°, and indeed the equation $5c^2-b^2=2d^2$ giving $c^2-d^2=\frac{b^2-3d^2}{5}$ is impossible by Art. 15; and if 3 were common measure we should have $c^2+b^2=3a^2$ which is impossible by Art. 14, 1°; and if 6 were CM we should have $c^2+b^2=6a^2$ which is impossible by Art. 14, 2°; lastly, if the fraction $\frac{c^2+b^2}{5c^2-b^2}$ were irreducible like $\frac{a^2}{d^2}$ we should have $c^2+b^2=a^2$ and $5c^2-b^2=d^2$ giving $6c^2=a^2+d^2$ which is impossible by Art. 14, 2°. Now taking the sign —, in the equations $m^2+n^2=z=\pm(5p^2-q^2)$ we get $a^2b^2+c^2d^2=b^2d^2-5a^2c^2$ or ∴ $\frac{b^2-c^2}{b^2+5c^2}=\frac{a^2}{d^2}$ and as before, the fraction $\frac{b^2-c^2}{b^2+5c^2}$ can have no common measure except 6 or a divisor of 6; now 2 cannot be CM, for if so we should have $b^2+5c^2=2d^2$ which is impossible by Art. 9, case 3°, or by Art. 15; nor could 3 be CM, for if so we should have $b^2+5c^2=3d^2$ which is impossible by Art. 15; and if 6 were CM we should have $b^2-c^2=6a^2$ and $b^2+5c^2=6d^2$, giving $c^2+a^2=d^2$ and $c^2+6a^2=b^2$ which being similar to the two original equations, but having c and a instead of x and y, and as c and a are $<x$ and y we ∴ infer that the proposed equations can admit of no solution in great integers since they admit of none in small integers; lastly, if $\frac{b^2-c^2}{b^2+5c^2}$ were irreducible like $\frac{a^2}{d^2}$ we should have $b^2-c^2=a^2$ and $b^2+5c^2=d^2$ or ∴ $a^2+c^2=b^2$ and

$a^2+6c^2=d^2$ which are also similar to the proposed equations, and from this their impossibility is inferred as in the preceding instance ∴ the proposed equations are incompatible. Q. E. D.

24.—The two simultaneous equations $\left\{\begin{matrix} x^2+y^2=\square=z^2 \\ x^2+8y^2=\square=w^2 \end{matrix}\right\}$ are impossible.

By elimination $7y^2=w^2-z^2$ and $7x^2=8z^2-w^2$ so that only one coeft. (8) contains a square factor. Now if the proposed equations be possible at all they must obviously be so when x is prime to y, and ∴ by Art. 5, x, y, z and w all prime to each other except perhaps x and w, which if not prime to each other, when the rest are so, could have no CM but 2; but this cannot be, for if x and w were even, y which is prime x should be odd, then putting $x=2x'$ and $w=2w'$ the 2nd equation gives $y^2=\frac{1}{2}(w'^2-x'^2)$ which is impossible since y^2 is odd, and $\frac{1}{2}(w'^2-x'^2)$ must clearly be even if it be at all an integer; thus y must be even and ∴ the other 3 odd, as they are prime to y, and then by Art. 1 the general solution of the first equation is $x=m^2-n^2$ and $y=2mn$, m being prime to n, even and odd; and by Art. 2 the general solution of the 2nd equation is $\pm x=2p^2-q^2$ and $y=pq$, q being odd and prime to p, which must be even since y is so; hence we must have $pq=2mn$, which, as q is odd, can subsist only thus: $q=ab$, $m=ac$, $n=bd$ and ∴ $p=2cd$, a, b, c, d being prime to each other, and one of them (c or d) even, else m and n would be both odd, then equating the values of x, first taking the sign $+$, viz. $m^2-n^2=2p^2-q^2$ gives $a^2c^2-b^2d^2=8c^2d^2-a^2b^2$ or ∴ $\frac{b^2+c^2}{b^2+8c^2}=\frac{d^2}{a^2}$ and, as in the preceding Articles, the left hand member can have no common measure but 7, since b is prime to c; nor can 7 be CM, for if so we should have $b^2+c^2=7d^2$ which is impossible by Art. 14, case 3°; and if there be no CM we should have $b^2+c^2=d^2$ and $b^2+8c^2=a^2$, which being exactly similar to the two proposed equations, but having b and c instead of x and y, and as b and c are $<x$ and y we ∴ infer as usual that the proposed equations can admit of no solution in great numbers as they admit of no solution in small integers; if we had taken $-x=2p^2-q^2$, then the equation $m^2-n^2=x=q^2-2p^2$ could be written $n^2-m^2=2p^2-q^2$ or ∴ $\frac{a^2+d^2}{a^2+8d^2}=\frac{c^2}{b^2}$ so that the conditions and demonstration would be exactly the same as before ∴, &c. Q. E. D.

Cor.—The equation $2x^4-3x^2+2=\square=z^2$ is impossible when x and z are rational: For putting $x^2-1=2y$ and $x^2+1=2w$ we get $x^2+y^2=w^2$ and $x^2+8y^2=z^2$, which were just proved to be impossible in integers, and ∴ also in fractions, see Art. 4, the case of $y=0$, or ∴ $x=\pm1$, being the only exception; see Art. 143 *Euler's Algebra.*

Cor. 2.—Hence also the two simultaneous equations, $x^2\pm2y^2=nz^2$ and $y^2\pm4x^2=nw^2$ are impossible, whatever rational number n may be; for by the general theorem, Art. 19, if these were possible, the equations, $x^2+y^2=\square$ and $x^2+8y^2=\square$ should be also possible; and by taking the lower signs $(-)$ and n negative, it follows that the equations $2y^2-x^2=nz^2$ and $4x^2-y^2=nw^2$ are also impossible, n being *any* rational number.

25.—The two simultaneous equations $\left\{\begin{matrix} x^2+y^2=\square=z^2 \\ x^2+9y^2=\square=w^2 \end{matrix}\right\}$ are impossible.

If these equations were possible at all, they should be so when x is prime to y, and not divisible by 3; for if θ were a common measure of x and y, we could divide off by θ^2; and if x were divisible by 3 and ∴ y not so divisible, since it is prime to x, then putting $3x'$ for x, we should have $y^2+x'^2=\square$ and $y^2+9x'^2=\square$ which are exactly similar to the original pair; but here y is *not* divisible by 3, hence then it will be sufficient to prove the impossibility when x is prime to y, and not divisible by 3 and $\therefore x$ prime to $3y$; now 1°— if x be even and $\therefore y$ odd, the general solution of the 1st equation is $x=2mn$ and $y=m^2-n^2$ m being prime to n, and they even and odd; and as x is prime to $3y$, and even $\therefore$ the general solution of the 2nd equation, by Art. 1, is $x=2pq$ and $3y=p^2-q^2$, hence $mn=pq$, which can subsist only thus: $m=ab$, $p=ac$, $q=bd$ and $\therefore n=cd$, a, b, c, d, being prime to each other, and one of them even, and then $p^2-q^2=3y \therefore =3(m^2-n^2)$ gives $a^2c^2-b^2d^2=3(a^2b^2-c^2d^2)$ or $\therefore$ $\frac{a^2+3d^2}{d^2+3a^2}=\frac{b^2}{c^2}$, now the fraction $\frac{a^2+3d^2}{d^2+3a^2}$ has numerator + denominator $=4(a^2+d^2)$ and denominator — numerator$=2(a^2-d^2)$ and as a is prime to d $\therefore a^2+d^2$ and a^2-d^2 can have no C M but 2, and that only when a and d are both odd $\therefore$ our fraction has no C M but 4 or a divisor of 4; if there be no C M we must have $a^2+3d^2=b^2$ and $d^2+3a^2=c^2$, giving $b^2+c^2=4(a^2+d^2)$ an impossible equation, since b^2+c^2 will have the form $4n+1$ or $4n+2$ according as one or both of b and c are odd: and they cannot be both even, as they are prime to each other; if 2 were C M we should have $a^2+3d^2=2b^2$ and $d^2+3a^2=2c^2$ each of which is impossible by Art. 15; and if 4 were C M we should have $a^2+3d^2=4b^2$ and $d^2+3a^2=4c^2$ giving $a^2+d^2=b^2+c^2$ which is obviously impossible, as only one of a, b, c, d, must be even, and the other three odd. 2°—if x be odd and $\therefore y$ even, then the general solution of the 1st equation is $x=m^2-n^2$ and $y=2mn$, and the general solution of the 2nd equation, when x is odd and prime to $3y$ is, by Art. 1, $x=p^2-q^2$ and $3y=2pq$ hence we must have $pq=3mn$, m being prime to n, and p to q, one of each pair being even, and the other odd; this last equation shows that either p or q must be divisible by 3; let it be p that is so, then the equation can subsist only thus: $p=3ab$, $m=ac$, $n=bd$, and $\therefore q=cd$, and then $m^2-n^2=x=p^2-q^2$ gives $a^2c^2-b^2d^2=9a^2b^2-c^2d^2$, if q, and not p were divisible by 3, we should find $a^2c^2-b^2d^2=a^2b^2-9c^2d^2$ or $\therefore$ $b^2d^2-a^2c^2=9c^2d^2-a^2b^2$, an equation exactly similar to that already found when p was supposed $\div 3$ so that it is quite sufficient to consider only this latter case of $p\div 3$ which gives $\frac{d^2+a^2}{d^2+9a^2}=\frac{b^2}{c^2}$ where a, b, c, d, are prime to each other, and one of them even, since m is prime to n, and p prime to q, and one of each pair even: now $\frac{d^2+a^2}{d^2+9a^2}$ can have no C M except 8 or a divisor of 8; if there were no C M we should have $d^2+a^2=b^2$ and $d^2+9a^2=c^2$, two equations precisely similar to the original pair, and thereby indicating that if there existed a solution in large integers, there should also exist a solution in smaller integers, and thence again in integers still smaller, &c., &c.; for $x=m^2-n^2=a^2c^2-b^2d^2=(ac+bd)(ac-bd)$ which is $>a, b, c$, or d, and $y=2mn \therefore =2abcd$ is also $> a, b, c$, or d; and so as there exists no solution in small integers, there can $\therefore$ can exist no solution whatever in integers nor even in fractions; see Arts. 4 and 5. If 2 were C M we should have $d^2+9a^2=2c^2$ or $\therefore 3a^2=\frac{2c^2-d^2}{3}$ which is impossible by Art. 15; it also gives $\frac{c^2+d^2}{3}=c^2-3a^2$ which is impossible by Art. 8, 1° or Art. 14, 1°. 4 or 8 cannot be common measure, since a is prime to d and $\therefore a^2+d^2$ of the form 4N+1 or 4N+2, according as only one, or both of a and d are odd $\therefore$ &c.

26.—The two simultaneous equations $\begin{Bmatrix} x^2+y^2=z^2 \\ x^2+12y^2=w^2 \end{Bmatrix}$ are impossible.

If these equations be possible, they must obviously be so when x is prime to y; and then x cannot be even; for if so y should be odd and ∴ by 1st equation z also odd, and as every odd square $=8n+1$ ∴ the first equation giving $x^2=z^2-y^2$ shows that x should be divisible by 4, let ∴ $x=4x'$ then $\frac{1}{4}(x^2+12y^2)=4x'^2+3y^2$ and as y is odd, this is of the form $4n+3$ and ∴ could not be a square, since *every* odd square has the form $4n+1$: Again if x be odd and y even, the general solution of the first equation is, by Art. 1, $x=m^2-n^2$, $y=2mn$; and by Art. 3, the general solution of the 2nd equation, when y is even, is $y=pq$ and $\pm x=3p^2-q^2$; m being prime to n, and p prime to q, one of each pair being even, and the other odd; hence $pq=2mn$, and if p (and not q) be even, this requires $p=2ab$, $m=ac$, $n=bd$, $q=cd$; a, b, c, d, being prime to each and one of them even, and as $m^2-n^2=x=\pm(3p^2-q^2)$ by taking the sign — gives us $n^2-m^2=3p^2-q^2$ which is quite similar to the condition or equation $m^2-n^2=3p^2-q^2$, got by taking the sign + (m and n being merely interchanged, which change does not affect the other condition $pq=2mn$) it will ∴ be sufficient to consider and disprove this latter equation, which gives $a^2c^2-b^2d^2=12a^2b^2-c^2d^2$ or $\frac{d^2+a^2}{d^2+12a^2}=\frac{b^2}{c^2}$ and as a is prime to d the left hand member could obviously have no common measure but 11, but by Art. 14. 4°, 11 could not divide d^2+a^2 when d is prime to a; so we should have $d^2+a^2=b^2$ and $d^2+12a=c^2$, which being similar to the proposed equations indicate thereby that there can be no solution in large integers, as there exists none in small integers. But if q (and not p) were even, then $pq=2mn$ could subsist only thus, $q=2ab$, $m=ac$, $n=bd$, $p=cd$, and then $m^2-n^2=x=+(3p^2-q^2)$ gives $a^2c^2-b^2d^2=3c^2d^2-4a^2b^2$ or $\frac{b^2+3c^2}{c^2+4b^2}=\frac{a^2}{d^2}$ and as before, the left hand member could have no common measure but 11, but this cannot be, since by Art. 15 each of the equations $b^2+3c^2=11a^2$ and $c^2+4b^2=11d^2$ is impossible; and if the left hand fraction were irreducible we should have $b^2+3c^2=a^2$ and $c^2+4b^2=d^2$ giving $(2b)^2+c^2=\square=d^2$ and $(2b)^2+12c^2=\square=(2a)^2$ which, being similar to the two original equations, thereby indicate the impossibility in great numbers, as no solution exists in small numbers.

Cor.—Hence by the general theorem, Art. 19, the two equations $\begin{Bmatrix} x^2\pm2y^2=n\,z^2 \\ y^2\pm6x^2=nw^2 \end{Bmatrix}$ are impossible, as also the pair $\begin{Bmatrix} x^2\pm3y^2=n\,z^2 \\ y^2\pm4x^2=nw^2 \end{Bmatrix}$

27. The two simultaneous equations $\begin{Bmatrix} x^2+y^2=\square=z^2 \\ x^2+13y^2=\square=w^2 \end{Bmatrix}$ are impossible.

If these two equations be possible at all they must obviously be so when x is prime to y, and as the first equation shows that x and y cannot be both odd, so the second equation shows that x could not be even and y odd, for if so x^2+13y^2 would be of the form $8n+5$, which can never be a square number, as every odd square$\equiv8n+1$, and this also shows that if x be even and ∴ y and z both odd in first equation, then x should be divisible by 4; so if the thing be possible at all x must be odd and y even, and then the general solution of the first equation is $x=m^2-n^2$, $y=2mn$, and by Art. 3, the general solution of the second equation under these circumstances is $y=2pq$ and $\pm x=13p^2-q^2$ so that we must have $mn=pq$ which can be satisfied only thus $m=ab$, $p=ac$,

$q=bd$ and $\therefore n=cd$: a, b, c, d, being prime to each other and one of them even, and then $m^2-n^2=x=+(13p^2-q^2)$ gives $a^2b^2-c^2d^2=13a^2c^2-b^2d^2$ or $\therefore$ $\frac{d^2+a^2}{d^2+13a^2}=\frac{c^2}{b^2}$; but if we take $-x=13p^2-q^2$ we would have $n^2-m^2=-x=13p^2-q^2$, an equation exactly similar to the foregoing, so that it is quite sufficient to prove the impossibility of $\frac{d^2+a^2}{d^2+13a^2}=\frac{c^2}{b^2}$, and here, if there be no common measure, we should have $d^2+a^2=c^2$ and $d^2+13a^2=b^2$ two equations precisely similar to the original pair, but having d and a (which are $<x$ and y) instead of x and y, &c., &c.

If 2 were CM we should have $d^2+13a^2=2b^2$ or $a^2=\frac{2b^2-d^2}{13}$ impossible by Art. 15. If 3 were CM we should have $d^2+a^2=3c^2$ which is impossible by Art. 14, 1°. If 4 were CM we should have $d^2+a^2=4c^2$ which is impossible when d is prime to a and $\therefore d^2+a^2=4n+1$ or $4n+2$. According as one or both of d and a are odd. 6 or 12 cannot be CM since d^2+a^2 is never divisible by 3 when d is prime a, Art. 8, 1°. Hence if it were possible at all it should also be possible in small integers, and as no solution exists in small numbers $\therefore$ the equations are impossible *simultaneously.*

28. The two simultaneous equations $\left\{\begin{matrix} x^2+y^2=\square=z^2 \\ x^2+14y^2=\square=w^2 \end{matrix}\right\}$ are impossible.

By subtraction $13y^2=w^2-z^2$ and here as no coeft. contains a square factor $\therefore$ by Art. 5, if the proposed eqnations be possible at all they should be so when x, y, z, and w are all prime to each other, and moreover x and w prime to the coeft. 14 and $\therefore$ both odd and $\therefore$ y even (by first equation) hence by Art. 1 the general solution of the first equation is $x=m^2-n^2$, $y=2mn$ and $z=m^2+n^2$ and by Art. 3 the general solution of $13y^2=w^2-z^2$, when y is even, is $y=2pq$ and $\pm z=13p^2-q^2$, m being prime to n, and p to q, one of each pair being even and the other odd, hence we must have $mn=pq$ which requires $m=ab$, $p=ac$ $q=bd$ and $\therefore n=cd$, a, b, c, d, being prime to each other, and one of them even, else m, n, p, and q would be all odd; and then $m^2+n^2=z=\pm(13p^2-q^2)$ gives, by taking the sign $+$, $a^2b^2+c^2d^2=13a^2c^2-b^2d^2$, or $\therefore$ $\frac{a^2+d^2}{13a^2-d^2}=\frac{c^2}{b^2}$ and if there be no CM we should have $a^2+d^2=c^2$ and $13a^2-d^2=b^2$ or $\therefore$ $14a^2=b^2+c^2$ which is impossible by Art. 14; if 2 were CM we should have $a^2+d^2=2c^2$ and $13a^2-d^2=2b^2$ or $\therefore 7a^2=b^2+c^2$ which is impossible by Art. 14, 3°. If 7 were CM we should have $a^2+d^2=7c^2$ which is also impossible by Art. 14, 3°. lastly if 14 were CM we should have $a^2+d^2=14c^2$ which is impossible by Art. 14, 5°. Now taking the sign—*i.e.* $z=-(13p^2-q^2)=q^2-13p^2$ then $m^2+n^2=z=q^2-13p^2$ gives $a^2b^2+c^2d^2=b^2d^2-13a^2c^2$ or $\therefore$ $\frac{d^2-a^2}{d^2+13a^2}=\frac{c^2}{b^2}$ and if there be no common measure we must have $d^2-a^2=c^2$ and $d^2+13a^2=b^2$ or $\therefore c^2+a^2=d^2$ and $c^2+14a^2=b^2$ two equations similar to the proposed pair, where c and a replace x and y and are $<x$ and y, showing that if there were a solution in great integers there should exist a solution in smaller, and thence in still smaller integers, &c. If 2 were CM we should have $d^2+13a^2=2b^2$ which is impossible by Art. 15. If 7 were CM we should have $d^2+13a^2=7b^2$ which is impossible by Art. 15. Lastly if 14 were CM we should have $d^2-a^2=14c^2$ and $d^2+13a^2=14b^2$ giving $a^2+c^2=b^2$ and $a^2+14c^2=d^2$

which are similar to the original pair of equations, as in the first case already considered $\therefore$ &c. Q.E.D.

Cor.—By carefully considering the final conditions, and the demonstration of Articles 23 and 28, we are naturally conducted to the discovery of the following

GENERAL THEOREM.

The solution of $X^2+Y^2=\square$ and $X^2+(a+1)Y^2=\square$ can be obtained from the solution of the two auxiliary equations $x^2\pm y^2=nz^2$ and $ay^2\mp x^2=nw^2$; in fact I say $X=x^2z^2-y^2w^2$ and $Y=2xyzw$ will answer.

DEMONSTRATION.

For then $X^2+Y^2=(x^2z^2+y^2w^2)^2$ and so the first condition is fulfilled; again $nX=x^2(x^2\pm y^2)-y^2(ay^2\mp x^2)=x^4\pm 2x^2y^2-ay^4$ and $n^2Y^2=4x^2y^2(x^2\pm y^2)(ay^2\mp x^2)=(4a-4)x^4y^4\mp 4x^6y^2\pm 4ay^6x^2$ and $\therefore n^2(X^2+\overline{a+1}.Y^2)=x^8\pm 4x^6y^2+(4-2a)x^4y^4\mp 4ax^2y^6+a^2y^8+(4a^2-4)x^4y^4\mp(4a+4)x^6y^2\pm(4a^2+4a)y^6x^2=x^8\mp 4ax^6y^2+(4a^2-2a)x^4y^4\pm 4a^2y^6x^2+a^2y^8=(x^4\mp 2ax^2y^2-ay^4)^2$ and $\therefore X^2+(a+1)Y^2=(x^2\mp 2ax^2y^2-ay^4)^2\div n^2\;\therefore=\square$ and thus the 2nd condition is also fulfilled.

Cor. 2.—From the preceding general Theorem and what has been demonstrated in Articles 17, 18, 22, 23, 24, 25, 26, 27, and 28, it follows that the two simultaneous equations $\begin{cases}x^2\pm y^2=nz^2\\ ay^2\mp x^2=nw^2\end{cases}$ are impossible whenever $a=1,2,3,4,5,7,8,11,12$ or 13, *whatever rational number n may be.*

29. The two simultaneous equations, $\begin{cases}x^2+y^2=\square=z^2\\ x^2+15y^2=\square=w^2\end{cases}$ are impossible.

Here $14y^2=w^2-z^2$ and as no coeft. contains a square factor $\therefore$ by Art. 5 if the proposed equations be possible at all they must be so when x, y, z, and w, are all prime to each other and also w and z prime to 14, and $\therefore$ both odd and $\therefore$ y even as the equation $14y^2=w^2-z^2$ indicates, as every odd square $=8N+1$; and y being even x, z, and w, which are prime to y must be odd; indeed it is evident that if x were even and y odd, x^2+15y^2 would be of the form $4n+3$, which can never be a square, as every odd square $=8N+1$; and now as y must be even $\therefore$ by Art. 1 the general solution of the 1st equation is, $x=m^2-n^2$, and $y=2mn$; and by Art. 3 the general solution of the 2nd. equation is $\pm x=15p^2-q^2$ or else $=5p^2-3q^2$ and $y=2pq$; m being prime to n and p prime to q one of each pair being even and the other odd; hence we must have $mn=pq$ which as usual requires $m=ab$, $p=ac$, $q=bd$, and $n=cd$, a, b, c, d, being prime to each other, and one of them even; $m^2-n^2=x=\pm(15p^2-q^2)$ and here taking the sign $-$, gives $n^2-m^2=15p^2-q^2$ which is exactly similar to $m^2-n^2=+(15p^2-q^2)$ got by taking the sign $+$, $\therefore$ it will suffice to prove the impossibility of this latter equation, which gives $a^2b^2-c^2d^2=15a^2c^2-b^2d^2$ or $\therefore\ \dfrac{d^2+a^2}{d^2+15a^2}=\dfrac{c^2}{b^2}$ and as a is prime to d it is evident $\dfrac{d^2+a^2}{d^2+15a^2}$ can have no common measure but 14 or a divisor of 14; if 14 were CM we should have $d^2+a^2=14c^2$ which is impossible by Art. 14. 5°—If 7 were CM we should have $d^2+a^2=7c^2$ which is impossible by Art. 14. 3°—If 2 were CM we should have $d^2+15a^2=2b^2$ or $3a^2=\dfrac{2b^2-d^2}{5}$ which is impossible by Art. 15, or it gives $\dfrac{b^2+d^2}{3}=b^2-5a^2$ which is impossible by Art. 8,1°. Lastly, if there be no CM we should have

$d^2+a^2=c^2$ and $d^2+15a^2=b^2$ two equations precisely similar to the original pair, and thereby proving that there exists no solution in great integers as there exists none in small integers.

But if we take $\pm x=5p^2-3q^2$, then for reasons above assigned, it will be sufficient to consider the sign $+$, as the other sign $(-)$ leads to a perfectly similar equation and condition; now $m^2-n^2=x=+(5p^2-3q^2)$ gives $a^2b^2-c^2d^2=5a^2c^2-b^2d^2$ or $\frac{a^2+3d^2}{d^2+5a^2}=\frac{c^2}{b^2}$ and the left hand member has 5.numerator—denominator$=14d^2$, and 3.denominator—numerator$=14a^2$ and as a is prime to d $\therefore$ if there be a common measure it must be 14 or a divisor of 14; if there be no common measure, we should have $a^2+3d^2=c^2$ and $d^2+5a^2=b^2$ giving $2a^2=\frac{3b^2-c^2}{7}$ which is impossible by Art, 15.—If 2 were common measure, we should have $a^2+3d^2=2c^2$ or $d^2=\frac{2c^2-a^2}{3}$ impossible by Art. 15; if 7 were C M we should have $d^2+5a^2=7b^2$ or $a^2-b^2=\frac{2b^2-d^2}{5}$ which is impossible by Art. 15; if 14 were C M we should have $a^2+3d^2=14c^2$ or $\frac{a^2+c^2}{3}=5c^2-d^2$ which is impossible by Art. 15, or Art. 14, 1° or Art 8, 1°. $\therefore$ &c. Q.E.D.

30.— The two simultaneous equations $\begin{cases} x^2+y^2=\square=z^2 \\ x^2+16y^2=\square=w^2 \end{cases}$ are impossible. If it were possible at all, it should be so when x is prime to y and $\therefore z$ odd and prime to x and y. Now 1° if x were odd, and $\therefore y$ even the general solution of the 1st equation would be, by Art. 1, $x=m^2-n^2, y=2mn$. m being prime to n and they even and odd, and since x is supposed odd and prime to y $\therefore x$ is prime to $4y$, and hence by Art 1, the general solution of the 2nd equation is $x=p^2-q^2$ and $4y=2pq$, hence then we must have $pq=4mn$, as either p or q must be odd, let it be p that is so and $\therefore$ prime to 4, then this last equation can subsist only thus, $p=ab$, $m=ac$, $n=bd$ and $\therefore q=4cd$; a,b,c,d, being prime to each other, and one of them even, and then $m^2-n^2=x^2=p^2-q^2$ gives $a^2c^2-b^2d=a^2b^2-16\,c^2d^2$, or $\therefore$ $\frac{a^2+d^2}{a^2+16d^2}=\frac{c^2}{b^2}$ (if q and not p were odd, we could then write $m^2-n^2=p^2-q^2$, thus viz., $n^2-m^2=q^2-p^2$, and so the conditions and the demonstration would be obviously the same.) and if the left hand member have a common measure, it must obviously be a divisor of 15; if 15 were C M we should have $a^2+d^2=15c^2$, which is impossible by Art. 14, 6°. If 5 were C M we should have $a^2+16d^2=5b^2$ or $16d^2=5b^2-a^2$ which is impossible, since if $5b^2-a^2$ be even at all, a and b must be both odd, and then $5b^2-a^2$ is of the form $8n'+4$, as every odd square has the form $8n'+1$.

3 could not be common measure since $a^2+d^2=3c^2$ is impossible by Art. 14, 1°; and if $\frac{a^2+d^2}{a^2+16d^2}$ be irreducible like $\frac{c^2}{b^2}$ we should have $a^2+d^2=c^2$ and $a^2+15d^2=b^2$, two equations similar to the original pair, and thereby indicating (since a and d are $<x$ and y) that if it were possible in great integers it should be so in smaller, and thence again in still smaller integers, &c., &c. 2°. If x were even, and $\therefore y$ and z both odd, then, as in Articles 26, 27, x should be divisible by 4 and $\therefore=4x'$, so that we should have $y^2+x'^2=\square$ and $y^2+16x'^2=\square$, y being odd and prime to x', but this was just proved to be possible. Q.E.D.

31. The two simultaneous equations $\left\{\begin{matrix} x^2+y^2=\square=z^2 \\ x^2+18y^2=\square=w^2 \end{matrix}\right\}$ are impossible.

Here x being prime to y and $\therefore z$ odd and prime to both, the second equation shows that x cannot be even and y odd, for if so, x^2+18y^2 would $=4n'+2$ and $\therefore$ could not be a square number; hence if the thing be possible at all, x must be odd and y even and $\therefore$ by Art. 1 the general solution of the first equation is $x=m^2-n^2$ and $y=2mn$, and by Art. 3, the general solution of the derived equation $17y^2=w^2-z^2$ is $y=2pq$ and $\pm z=17p^2-q^2$, m being prime to n, and p to q, one of each pair being even, and the other odd, hence we must have $mn=pq$ which requires $m=ab$, $p=ac$, $q=bd$, and $\therefore n=cd$; a,b,c,d, being prime to each other and one of them even, as otherwise m,n,p and q would be all odd; with these values $m^2+n^2=z=\pm(17p^2-q^2)$ becomes by taking the sign $+$, $a^2b^2+c^2d^2=17a^2c^2-b^2d^2$, or $\frac{a^2+d^2}{17a^2-d^2}=\frac{c^2}{b^2}$ and if there be a common measure it must be a divisor of 18; if there be no C M we must have $a^2+d^2=c^2$ and $17a^2-d^2=b^2$ or $\therefore$ $\frac{b^2+c^2}{3}=6a^2$ which is impossible by Art 8, 1° or Art. 14, 1°.—If 2 were CM we should have $a^2+d^2=2c^2$ and $17a^2-d^2=2b^2$ or $\therefore$ $\frac{b^2+c^2}{3}=3a^2$ which is impossible for the same reason; 3, 6, 9, or 18 cannot be C M since a^2+d^2 is never divisible by 3 when a is prime to d, 8, 1°. Now by taking the sign $-$, the equation $m^2+n^2=z=-(17p^2-q^2)$ *i.e.* $=q^2-17p^2$ gives $a^2b^2+c^2d^2=b^2d^2-17a^2c^2$ or $\frac{d^2-a^2}{d^2+17a^2}=\frac{c^2}{b^2}$ and if there be no C M we shall have $d^2-a^2=c^2$ and $d^2+17a^2=b^2$ giving $c^2+a^2=d^2$ and $c^2+18a^2=b^2$ two equation similar to the original pair, ond thereby proving, since c and a are $<x$ and y, that the proposed equations could not have a solution in large integers without having also a solution in smaller, and thence again in still smaller integers, &c., &c. If 2 were C M we should have $d^2-a^2=2c^2$ and $d^2+17a^2=2b^2$ giving $b^2-c^2=9a^2$, $b^2+17c^2=9d^2$ or $(3a)^2+c^2=b^2$ and $(3a)^2+18c^2=(3d)^2$ which are similar to the proposed equations, but having x and y replaced by $3a$ and c which are $<x$ and y since $y=2mn$ $\therefore$ $=2abcd$ is $>c$ and $x=m^2-n^2=a^2b^2-c^2d^2=(ab+cd)(ab-cd)$ is $>3a^2$ as the equartion $b^2=(3a)^2+c^2$ proves $b>3a$ &c., &c. If 3 were C M, we should have $d^2+17a^2=3b^2$ or $a^2=\frac{3b^2-d^2}{17}$ which is impossible by Art. 15. If 6 were C M we should have $d^2+17a^2=6b^2$, which is also impossible by Art 15. If 9 were C M, we should have $d^2-a^2=9c^2$ and $d^2+17a^2=9b^2$, which returns to the last case but one already disproved; lastly, if 18 were C M we should have $d^2-a^2=18c^2$ and $d^2+17a^2=18b^2$ giving $a^2+c^2=b^2$ and $a^2+18c^2=d^2$ two equations again similar to the proposed equations, and thereby proving that there exists no solution in large integers, as none exist in small integers; and hence the two proposed equations are impossible *simultaneously*.

The two simultaneous equations $\left\{\begin{matrix} x^2+y^2=z^2 \\ x^2+19y^2=w^2 \end{matrix}\right\}$ are easily proved impossible by a method like that given in Art. 17; the first equation shows that x and y cannot be both odd (see Art. 1,) and the second equation shows that x could not be even and y odd, for if so, x^2+19y^2 would be of the form $4n+3$ which can never be a square, as *every* odd square has the form $8n+1$, &c., &c.

Cor.—From the general theorem in Art. 19 in connexion with what has been demonstrated in Articles 28, 29, 30, and 31, it follows that the pair of equations $\begin{cases} x^2 \pm ay^2 = nz^2 \\ y^2 \pm bx^2 = nw^2 \end{cases}$ is impossible when $ab=14$, 15, 16, 18 or 19 whatever rational numbers a and n may be; it follows also from the same Articles and the general theorem of Art. 28, that the pair of equations $x^2 \pm y^2 = nz^2$ and $ay^2 \mp x^2 = nw^2$ are impossible, *whatever rational number n may be* when $a=14$, 15, 17, or 18.

Cor. 2.—Since it is now *demonstrated* that $x^2+y^2=\square$ and $x^2+ay^2=\square$ are impossible when a is any integer <20 except 7, 10, 11, or 17; if we take $x=z^2-1$ and $y=2z$, then $x^2+y^2=(z^2+1)^2 \therefore =\square$ is fulfilled, and $\therefore x^2+ay^2$ cannot then become a square, and thus $z^4+(4a-2)z^2+1$ can never become a square when a is any integer between 1 and 20 except 7, 10, 11 or 17; which includes the expressions z^4+6z^2+1, z^4+10z^2+1, z^4+14z^2+1, z^4+18z^2+1, z^4+22z^2+1, z^4+30z^2+1, z^4+34z^2+1, &c., &c. And as it will be proved in some of the subsequent Articles that $x^2-y^2=\square$ and $x^2-ay^2=\square$ are impossible *simultaneously* when a is any integer between 1 and 18 except 7 or 11; it follows also, by taking $x=z^2+1$ and $y=2z$ that $z^4-(4a-2)z^2+1=\square$ is impossible when a is any integer between 1 and 13 except 7 or 11; this formula includes the expressions z^4-6z^2+1, z^4-10z^2+1, z^4-14z^2+1, z^4-18z^2+1, z^4-22z^2+1, z^4-30z^2+1, &c., &c.

SCHOLIUM.

As a great number of pairs of equations of the form $ax^2+by^2=nx^2$, and $cx^2-dy^2=nw^2$ are proved impossible simultaneously, *whatever rational number n may be*, in, or by means of Arts. 8 cor 2, 10, 12 cor 1, 19, 23, 24 cor 2, 26 cor, 28 cor 2, 31 cor 1, 36, 37, 39, 40, 41 and its cor, &c.; it follows $\therefore$ that we shall thence have a corresponding vast number of products $(ax^2+by^2)(cx^2-dy^2)$ none of which can ever be $=\pm\square$ *Ex, gr.* since by chap. 1, and especially Art. 10, $ax^2+by^2=nz^2$ and $ax^2-by^2=\pm nw^2$ are impossible *simultaneously, whatever rational n⁰ n may be*, when ab is any integer <20 except 5, 6, 7, 13, 14, or 15, it follows $\therefore$ that $(ax^2+by^2)(ax^2-by^2)$ or $a^2x^4-b^2y^4$ can never $=\pm\square$ when ab is any integer <20 except those above mentioned; thus neither x^4-y^4 nor x^4-4y^4 nor x^4-9y^4 nor x^4-100y^4 nor $4x^4-25y^4$ &c., can ever be $=\pm\square$. And from Art. 11 it follows that $n(x^4-y^4)=\square$ is impossible when n is any integer <20 except those above mentioned, and hence by putting $x=p^2+mq^2$ and $y=p^2-mq^2$ and $\therefore m(x^2-y^2)=\square$, it follows that $nm(x^2+y^2)=2n(mp^4+m^3q^4)=\square$ is always impossible, n being the same as before; thus $m=n$ proves the impossibility of $2(p^4+n^2q^4)=\square$; n as stated already, being any integer <20 except 5, 6, 7, 13, 14, or 15. Again by Art. 19, and the rest of this present chapter $x^2+ay^2=nz^2$ and $y^2+bx^2=nw^2$ are impossible *simultaneously*, when ab is any integer between 1 and 20 except 7, 10, 11 or 17, n being *any* rational number; it follows $\therefore$ that $(x^2+ay^2)(y^2+bx^2)=\square$ is *always impossible* when ab is restricted by the above limitation; but in other respects a and b may be *any numbers whatever* of like signs, and one of them may, of course, be $=\pm1$.

It will appear in like manner, from Art. 36, and chap. 3, that $(x^2-ay^2)(bx^2-y^2)=\square$ is *always impossible* when ab is any integer between 1 and 18, except 7 or 11. And from Art. 28 cor 1 and 2, it follows that $(x^2\pm y^2)(ay^2\mp x^2)=\square$ is impossible when a is any integer <19 except 6, 9, 10 or 16;

lastly, it follows from Art. 41, that neither $(x^2-y^2)(ax^2+y^2)$ nor $(x^2+y^2)(x^2-ay^2)$ can $=\Box$ when a is any integer <17 except 6 or 10.

Again, I say, $x^4+a^2y^4=\Box=z^2$ is impossible when $2a$ is any integer <20 except 5, 6, 7, 13, 14 or 15; For it gives $z^2\pm 2ax^2y^2=\Box$ which by chap. 1, are impossible *simultaneously* in the other cases: hence then, neither x^4+y^4 nor x^4+4y^4 nor x^4+25y^4 nor x^4+36y^4 nor $4x^4+9y^4$ nor $4x^4+121y^4$ &c., &c., can ever be $=\Box$ and thus we have a very considerable addition to the few cases of this kind heretofore recorded and proved; see *Euler's Algebra*, part 2, chap. 13.

From the known impossibility of $x^4+ky^4=\Box$ we can deduce as follows the impossibility of other similar formulæ. For as x^4+ky^4 cannot $=(x^2+\frac{b}{a}y^2)^2$ whatever rational numbers a and b may be $\therefore$ by reduction $\frac{x^2}{y^2}=\frac{ka^2-b^2}{2ab}$ is impossible, i. e. $\frac{ka^2-b^2}{2ab}$ can never be a square number, and $\therefore 2ab(ka^2-b^2)$ cannot become a square number. Now let $a=my^2$ and $b=\pm nx^2$ $\therefore$ $2mn(km^2y^4-n^2x^4)=\pm\Box$ is impossible; $m=2$ and $n=1$ give $x^4-4ky^4=\pm\Box$ impossible; $m=1=n$ gives $2(x^4-ky^4)=\pm\Box$ impossible. *Ex. gr.* $x^4+2y^4=\Box$ is proved impossible (*Euler*, Art. 210) $\therefore$ by what is here proved $2(x^4-2y^4)=\pm\Box$ is impossible; but its half, viz.: $x^4-2y^4=\Box$ is possible. See *Euler's Algebra*, Art. 211. It is easy to see that $x^4+kx^2y^2+y^4=\Box$ is always *possible* if $k=\pm 2$ or $-n^2$, or $2n^2\pm 2$, or $2n^2\pm 6$, or $2n^2\pm 34$, or $3n^2\pm 2$, or $3n^2\pm 4$, or $3n^2\pm 14$, or $5n^2\pm A$, if $A=2$, 3, 7, 18, 47, &c., or $6n^2\pm A$, if $A=2$, 10, 98, &c., or $7n^2\pm A$, if $A=2$, 16, 254, &c., or $21n^2\pm 110$, &c., &c., since $x^4+kx^2y^2+y^4=z^2$ is obviously fulfilled if we take $a^2-4=mb^2$, $k=mc^2\pm a$, $x=b$, $y=2c$ and $z=b^2\pm 2ac^2$. But any of these solutions that give $x=y$ or else x (or y)$=0$ would not lead to other solutions; thus $k=2n^2+6$ is $=14$ if $n=2$; and yet the equation $x^4+14x^2y^2+y^4=\Box$ or, (dividing off by y^4), $x^4+14x^2+1=\Box$ has, *undoubtedly*, only one solution, viz., $x=1$ as stated in the preceding Cor. 2: the equation is also *possible* when k is negative and equal to any of the following integers, viz., 2, 4, 9, 11, 13, 15, 16, 25, 26, 27, 28, 32, 36, 39, 40 42, 43, 44, 47, 49, 51, &c., &c.

Chapter III.

On the Possible and Impossible cases of the two Simultaneous Equations.
$x^2-y^2=\square$ and $x^2-ay^2=\square$.

32.—The two simultaneous equations $\begin{Bmatrix} x^2-y^2=\square=z^2 \\ x^2-3y^2=\square=w^2 \end{Bmatrix}$ are impossible. By subtraction $2y^2=z^2-w^2$, and as no coeft. of these equations contains a square factor $\therefore$ by Art. 5, if the proposed equations be possible at all they must be so too when x, y, z an w are all prime to each other, and also z and w prime to coeft. 2 (of $2y^2$) and $\therefore$ both odd and $\therefore$ y even, as the equation $2y^2=z^2-w^2$ then proves, and y being even x, z and w which are prime to y, must be odd; hence, by Art. 1, the general solution of the 1st equation is $x=m^2+n^2$ and $y=2mn$; and, by Art. 3, the general solution of the 2nd equation, when y is even, is $x=3p^2+q^2$ and $y=2pq$; m being prime to n, and p to q, one of each pair being even and the other odd; thus we must have $mn=pq$ which requires $m=ab$, $p=ac$, $q=bd$, and $\therefore$ $n=cd$, $abcd$ being prime to each other, and one of them even; and then $m^2+n^2=x=3p^2+q^2$ gives $a^2b^2+c^2d^2=3a^2c^2+b^2d^2$ or $\dfrac{d^2-a^2}{d^2-3a^2}=\dfrac{c^2}{b^2}$, Now the left hand member has numerator—denominator$=2a^2$ and3 times numerator—denominator$=2d^2$,and as a is prime to d $\therefore$ there can be no common measure but ± 2; nor can $+2$ be CM, for if so we should have $d^2-3a^2=2b^2$ or $a^2=\dfrac{b^2-2d^2}{3}$ which is impossible by Art. 15; and if -2 were CM we should have $d^2-a^2=-2c^2$ and $d^2-3a^2=-2b^2$, giving $b^2-c^2=a^2$ and $b^2-3c^2=d^2$ two equations similar to the proposed, and $\therefore$ as usual, establishing the impossibility of the proposed equations, as there exists no solution in small integers; if there were no CM we should have $d^2-a^2=\pm c^2$ and $d^2-3a^2=\pm b^2$; taking the upper signs would give two equations similar to the proposed, as in the foregoing case, and taking the lower signs, (—) would give $b^2-c^2=2a^2$ and $b^2-3c^2=2d^2$; but this last equation is impossible by Art. 15, as already observed in the 1st case, $\therefore$ &c. Q. E. D.

33.—The two simultaneous equations $\begin{Bmatrix} x^2-y^2=\square=z^2 \\ x^2-4y^2=\square=w^2 \end{Bmatrix}$ are impossible. If the proposed equations were possible they should obviously be so when x is prime to y, and $\therefore$ by 1st equation x, y, z all prime to each other, and x odd; and then y could not also be odd, for if so x^2-4y^2 would be of the form $8n+5$, which can never be a square, as *every* odd square has the form $8n+1$, and now as x is odd, and y even, and x prime to y or to $2y$ $\therefore$ by Art. 1, the general solution of the first equation is $x=m^2+n^2$ and $y=2mn$; and the general solution of the 2nd equation is, by Art. 1, $x=p^2+q^2$ and $2y=2pq$ $\therefore$ we must have $pq=2mn$, m being prime to n, and p to q, one of each pair being even and the other odd. Now supposing p odd, this last equation can subsist only thus: $p=ab$, $m=ac$, $n=bd$, and $\therefore$ $q=2cd$, a, b, c, d being prime to each other, and one of them (not a or b) even, else m and n would be both odd; then $m^2+n^2=x=p^2+q^2$ gives $a^2c^2+b^2d^2=a^2b^2+4c^2d^2$ or $\therefore$ $\dfrac{a^2-d^2}{a^2-4d^2}=\dfrac{c^2}{b^2}$ and if there be no CM we must have $a^2-d^2=\pm c^2$ and $a^2-4d^2=\pm b^2$

and if there be a CM it can only be 3, and ∴ we should have $a^2-d^2=\pm 3c^2$ and $a^2-4d^2=\pm 3b^2$ which give $b^2-c^2=\mp d^2$ and $b^2-4c^2=\mp a^2$, which system or pair of equations is entirely similar to those obtained in the first case, so that it will be quite sufficient to prove the impossibility in the 1st case. Now if the signs be + in the 1st case, we have $a^2-d^2=c^2$ and $a^2-4d^2=b^2$ a pair of equations exactly similar to the original pair, and from which we infer the impossibility by the usual argument; but if the signs be —, we get $a^2-d^2=-c^2$ and $a^2-4d^2=-b^2$, this last gives $a^2+b^2=4d^2$ which is impossible, because when a is prime to b, a^2+b^2 will be of the form $4n'+1$ or $4n'+2$ according as one, or both of a and b are odd, ∴ &c. Q. E. D.

Cor.—Hence, by the method of Cor. 2, Art. 12, each of the following pairs is impossible, viz., $\left|\begin{matrix} z^2+y^2=x^2 \\ z^2-3y^2=w^2\end{matrix}\right|\begin{matrix} x^2-w^2=\square \\ x^2+3w^2=\square\end{matrix}\ \left|\begin{matrix} w^2+3y^2=z^2 \\ w^2+4y^2=x^2\end{matrix}\right|\begin{matrix} x^2-z^2=y^2 \\ 4z^2-3x^2=w^2\end{matrix}$ $\left|\begin{matrix} z^2-w^2=3y^2 \\ 4z^2-w^2=3x^2\end{matrix}\right|$; and also $\left|\begin{matrix} x^2-y^2=z^2 \\ x^2-4y^2=-w^2\end{matrix}\right|$ are impossible, as they give by subtraction, $3y^2=z^2+w^2$ which impossible by Art 14, 1°; so also $\left\{\begin{matrix} x^2+y^2=w^2 \\ 2x^2-y^2=z^2\end{matrix}\right\}$ are impossible by Art. 14, 1°, as they too give $3x^2=z^2+w^2$; and as $\left\{\begin{matrix} x^2-y^2=\square \\ x^2-4y^2=\square\end{matrix}\right\}$ are proved impossible ∴ they must be impossible when x is even and prime to y, changing ∴ x into $2x$ we see that $x^2-y^2=\square$ and $4x^2-y^2=\square$ are impossible simultaneously, &c., &c.

34.—The two simultaneous equations $\left\{\begin{matrix} x^2-y^2=z^2 \\ x^2-5y^2=w^2\end{matrix}\right\}$ are impossible. Here $4y^2=z^2-w^2$, and ∴ by Art. 5, if the proposed equations be possible they must be so too when x, y, z, w are all prime to each other, except, perhaps, z and w, which if not prime to each other, when the rest are so, can have no CM but 2; if they have this common measure, and are ∴ both even, then x and y, which are prime to them, and to each other, must be odd, and then by Articles 1 and 3, the general solution of the two equations is $x=m^2+n^2$, $y=m^2-n^2$, and $x=\frac{1}{2}(5p^2+q^2)$, $y=pq$, m being prime to n, one even and the other odd; and p prime to q, and both odd, thus, we should have $pq=m^2-n^2=(m+n)(m-n)$ which can be satisfied only thus, viz., $p=ab$, $m+n=ac$, $m-n=bd$ and $q=cd$, a, b, c, d being prime to each other, and all odd; then the other condition, viz., $m^2+n^2=x=\frac{1}{2}(5p^2+q^2)$ gives $(m+n)^2+(m-n)^2=5p^2+q^2$, i. e. $a^2c^2+b^2d^2=5a^2b^2+c^2d^2$ or ∴ $\dfrac{d^2-a^2}{d^2-5a^2}=\dfrac{b^2}{c^2}$ and if there be a common measure, it can only be 2 or 4; if there be no CM we must have $d^2-a^2=\pm b^2$ and $d^2-5a^2=\pm c^2$, when we take the signs +, the equations are similar to the proposed pair which are then proved impossible by the usual argument; if we take the signs —, then $c^2-b^2=(2a)^2$ and $c^2-5b^2=(2d)^2$ which are also similar to the proposed equations, &c.; it is evident 2 could not be CM as the equation $d^2-5a^2=\pm 2c^2$ is impossible by Art. 15; and if 4 were CM we should return again to the 1st case, only using $2b$ and $2c$ instead of b and c. Now if z be odd, and ∴ by Art. 1, y even, and then x, z, and w, which are prime to y odd; then, by Articles 1 and 3, the solutions of the two equations are $x=m^2+n^2$, $y=2mn$, and $x=5p^2+q^2$, $y=2pq$; m being prime to n, and p to q, one of each pair being even and the other odd, so we must have $mn=pq$ which requires $p=ab$, $m=ac$, $n=bd$, and $q=cd$, $abcd$ being prime to each other, and one of them even, and then the other condition $m^2+n^2=5p^2+q^2$ gives $a^2c^2+b^2d^2=5a^2b^2+c^2d$ or $\dfrac{d^2-a^2}{d^2-5a^2}=\dfrac{b^2}{c^2}$ which has been already discussed and disproved, ∴ &c. Q. E. D.

35.—The two simultaneous equations $\left\{\begin{array}{l} x^2-y^2=z^2 \\ x^2-6y=w^2 \end{array}\right\}$ are impossible. Here, $5y^2=z^2-w^2$, and as no coeft. contains a square factor $\therefore$ by Art. 5, if these two equations be possible they must be so when x, y, z, and w are all prime to each other, and also x and w prime to the coeft. 6 and $\therefore$ both odd, and then the 2nd equation shows, that y should be even and $\therefore$ the other 3 odd, as being prime to y; hence, by Art. 1, the general solution of the 1st equation is $x=m^2+n^2$, $y=2mn$, and $z=m^2-n^2$; and, by Art. 3, the general solution of $5y^2=z^2-w^2$ is $y=2pq$ and $z=5p^2+q^2$, m being prime to n, and p to q, one of each pair being even and the other odd; hence we must have $mn=pq$ which requires $m=ab$, $p=ac$, $q=bd$, and $n=cd$, a, b, c, d, being prime to each other, and one of them even; and then the 2nd condition $m^2-n^2=z=5p^2+q^2$ gives $a^2b^2-c^2d^2=5a^2c^2+b^2d^2$ or $\dfrac{a^2-d^2}{5a^2+d^2}=\dfrac{c^2}{b^2}$ (or $\dfrac{b^2+c^2}{b^2-5c^2}=\dfrac{a^2}{d^2}$) and if there be a common measure it must obviously be a divisor of 6; if there be no CM we should have $a^2-d^2=c^2$ and $5a^2+d^2=b^2$ giving $6a^2=b^2+c^2$, which is impossible by Art. 14, 2°. If 2 were CM we should have $5a^2+d^2=2b^2$ which is impossible by Art. 15. If 3 were CM we should have $5a^2+d^2=3b^2$ which is also impossible by Art. 15: and if 6 were CM we should have $a^2-d^2=6c^2$ and $5a^2+d^2=6b^2$, giving $a^2-c^2=b^2$ and $a^2-6c^2=d^2$ a pair of equations exactly similar to the original pair, having a and c instead of x and y; but $x=m^2+n^2=a^2b^2+c^2d^2$ is $>a$, and $y=2mn=2abcd$ is $>c$, so that if there existed a solution of the proposed equations in large integers there should also exist a solution in smaller integers, and thence in integers still smaller, &c., &c.; and as there exists no solution in small integers $\therefore$ there exists no solution whatever. See Article 4.

36.—From observing that the equation $\dfrac{d^2-a^2}{d^2-3a^2}=\dfrac{c^2}{b^2}$ in Art. 32 would be fulfilled if we could have $d^2-a^2=nc^2$ and $d^2-3a^2=nb^2$, and that we would then have $x=a^2b^2+c^2d^2$ and $y=2abcd$ to answer the equations $x^2-y^2=\square=z^2$ and $x^2-3y^2=\square=w^2$, and from a similar remark on the final equation or condition $\dfrac{d^2-a^2}{d^2-4a^2}=\dfrac{c^2}{b^2}$ in Art. 33, where we would then have $x=a^2b^2+c^2d^2$ $y=2abcd$ to fulfil $x^2-y^2=\square$ and $x^2-4y^2=\square$; and from the like remark on the final equation of Art. 34 we are naturally conducted to the discovery of the following

GENERAL THEOREM.

The values of X and Y, to fulfil $X^2-Y^2=\square=Z^2$ and $X^2-abY^2=\square=W^2$ can be obtained from the solution of the *auxiliary* equations $x^2-ay^2=nz^2$, $bx^2-y^2=nw^2$: in fact I say $X=x^2w^2+y^2z^2$ and $Y=2xyzw$ will answer

DEMONSTRATION.

For then $X^2-Y^2=(x^2w^2-y^2z^2)^2$, and so the first condition is obviously fulfilled. Again $nX=x^2(bx^2-y^2)+y^2(x^2-ay^2)=bx^4-ay^4$ and $n^2Y^2=4x^2y^2(nz^2)(nw^2)$ $=4x^2y^2(x^2-ay^2)(bx^2-y^2)=4bx^6y^2-4x^4y^4(1+ab)+4ax^2y^6$ and $\therefore n^2(X^2-abY^2)=b^2x^8-2abx^4y^4+a^2y^8-4ab^2x^6y^2+4x^4y^4(ab+a^2b^2)-4a^2bx^2y^6$ $\therefore=$ $(bx^4-2abx^2y^2+ay^4)^2$ and $\therefore X^2-abY^2=(bx^4-2abx^2y^2+ay^4)^2\div n^2$ $\therefore=\square$, and thus the 2nd condition is also fulfilled. Q. E. D.

Cor.—By taking $b=1$ and interchanging z and w in this general Theorem, we see that the solution of $X^2-Y^2=Z^2$ and $X^2-aY^2=W^2$ can be obtained from the solution of $x^2-y^2=nz^2$ and $x^2-ay^2=nw^2$, merely by taking $X=x^2z^2+y^2w^2$ and $Y=2xyzw$; and then again by taking $n=1$, this general theorem shows us how to find a solution in great integers from a known solution in smaller integers, of $x^2-y^2=z^2$ and $x^2-ay^2=w^2$; for then new $X=x^2z^2+y^2w^2$ $\therefore =x^4-ay^4$ and new $Y=2xyzw$ in all cases.

Ex. gr.—Let $a=7$ so that the two equations to be solved are $x^2-y^2=\square=z^2$ and $x^2-7y^2=\square=w^2$, then taking $n=2$, viz., a *possible* remainder of squares to modulus or divisor 7, we see that one obvious solution of the two auxiliary equations $x^2-y^2=2z^2$ and $x^2-7y^2=2w^2$ is $x=3$, $y=1$, $z=2$, and $w=1$. (Indeed the 1st equation shows, by Art, 5, that in a *primitive* solution, x and y should be both odd and z even); and $\therefore$ by the preceding Corollary, $X=x^2z^2+y^2w^2=37$ and $Y=2xyzw=12$ which are the *least* integers to answer the two proposed equations, giving $z=35$ and $w=19$; and from this solution we find another as indicated above, viz., new $X=x^4-ay^4=37^4-7.12^4=1729009$ and new $Y=2xyzw=37\times24\times35\times19=590520$, and now using these values of X and Y for x and y, we thence get another solution by the same formulae, viz. new $X=x^4-ay^4=1729009^4-7.590520^4=$, &c. . . . . . . As another example let $a=11$ so that the two equations to be solved are $x^2-y^2=\square=z^2$ and $x^2-11y^2=\square=w^2$; then taking $n=5$, viz., a possible remainder of squares to modulus 11 we see that one obvious solution of the two *auxiliary* equations $x^2-y^2=5z^2$ and $z^2-11y^2=5w^2$ is $x=7, y=2, z=3$ and $w=1$. (Indeed the derived equation $z^2-w^2=2y^2$ shows, by Art. 5, that z and w should be both odd, and y even, in a *primitive* solution,, and we $\therefore$ try $w=1$, $y=2$, $z=3$,) and $\therefore$ by the foregoing Cor. $X=x^2z^2+y^2w^2=21^2+2^2=445$, and $Y=2xyxw=84$ and these are the *least* integral values of x and y in the proposed equations, they give $z=437$ and $w=347$; and now from this solution we find another as indicated above, viz., new $X=x^4-ay^4=445^4-11.84^4$ and new $Y=2xyxw=2\times445\times84\times437\times347=$ &c.; and by using these numbers for x and y we can thence again find X and Y in very great integers, &c.

37.—The two simultaneous equations $\begin{Bmatrix} x^2-y^2=z^2 \\ x^2-8y^2=w^2 \end{Bmatrix}$ are impossible.

Here $7y^2=z^2-w^2$ and as no coeft. but 8 contains a square factor $\therefore$ by Art. 5, if the proposed equations were possible, they should be so too when x, y, z and w are all prime to each other, except, perhaps, x and w, which could have no CM but 2, when the rest are prime to each other; but even this exception has not place, since by Art. 1, x must be odd; now 1°— if y be even and $\therefore$ the other three, x, z and w odd, as being prime to y, the general solution of the 1st equation will be, by Art. 1, $x=m^2+n^2$, $y=2mn$, $z=m^2-n^2$, and by Art. 3, the general solution of the *derived* equation $7y^2=z^2-w^2$ will be $y=2pq$ and $z=7p^2+q^2$, m being prime to n, and p to q, one of each pair being even and the other odd, and then $mn=\frac{1}{2}y=pq$ requires $m=ab$, $p=ac$, $q=bd$ and $n=cd$, a, b, c, d, being prime to each other, and one of them even; and then the 2nd condition, $m^2-n^2=z=7p^2+q^2$ gives $a^2b^2-c^2d^2=7a^2c^2+b^2d^2$ or $\frac{a^2-d^2}{7a^2+d^2}=\frac{c^2}{b^2}$ (or $\frac{b^2+c^2}{b^2-7c^2}=\frac{a^2}{d^2}$) and it is plain if there be a common measure it must be a divisor of 8; if there be no CM we should have $a^2-d^2=c^2$ and $7a^2+d^2=b^2$ giving $8a^2=b^2+c^2$ which is impossible; for as b is prime to c, b^2+c^2 will be of the form $4n'+1$ or $8n'+2$ according as one or both of b and c

are odd; it is evident 2 or 4 could not be CM, for if so, a and d should be both odd, and b and c both even, but this cannot be, as b is prime to c and $\therefore$ only one of them (at most) even; lastly, if 8 were CM, we should have $a^2-d^2=8c^2$ and $7a^2+d^2=8b^2$ giving $a^2-c^2=b^2$ and $a^2-8c^2=d^2$ two equations exactly similar to the proposed, and $\therefore$ proving these latter impossible, as there exists no solution in small integers.

2°—If y be odd and $\therefore$ by Art. 1, z even, and the other three x, y, w odd as being prime to z, then by Art. 1, the general solution of the 1st equation is $x=m^2+n^2$, $y=m^2-n^2$ and $z=2mn$; and by Art. 3, the general solution of the derived equation $7y^2=z^2-w^2$ when y is odd, is $y=pq$ and $z=\frac{1}{2}(7p^2+q^2)$, m being prime to n, one even and the other odd, and p prime to q, and both odd, and then $m^2-n^2=y=pq$ can subsist only thus, viz. $m+n=ab$, $p=ac$, $q=bd$ and $m-n=cd$; a, b, c, d being prime to each other, and all odd, and then the 2nd condition $2mn=z=\frac{1}{2}(7p^2+q^2)$ or $(m+n)^2-(m-n)^2=7p^2+q^2$ gives $a^2b^2-c^2d^2=7a^2c^2+b^2d^2$ or $\dfrac{a^2-d^2}{7a^2+d^2}=\dfrac{c^2}{b^2}$ which is the very equation already discussed and disproved, Q. E. D.

38.—The two simultaneous equations, $\left\{\begin{matrix} x^2-y^2=z^2 \\ x^2-9y^2=w^2 \end{matrix}\right\}$ are impossible.

If the proposed equations be possible at all, they must obviously be so when x is prime to y and $\therefore$ ℣ 1st equation, x, y, z, all prime to each other and x odd; the two derived equations $8y^2=z^2-w^2$ and $8x^2=9z^2-w^2$ show that z and w must *then* be either prime to each other, or have no common measure but 2; now they cannot have even this common measure, for if so, y which is prime to z, should be odd, and putting $z=2z'$ and $w=2w'$ the equation $8y^2=z^2-w^2$ gives $y^2=\frac{1}{2}(z'^2-w'^2)$ which is impossible as y^2 is odd, and $\frac{1}{2}(z'^2-w'^2)$ will obviously be even if it be at all an integer; thus then w is prime to z and both odd and $\therefore$ y even, ℣ 1st equation, whose general solution is then $x=m^2+n^2$, $y=2mn$, $z=m^2-n^2$ and by Art. 2, the general solution of $8y^2=z^2-w^2$ is $y=pq$, $z=2p^2+q^2$, q being odd and prime to p, which must be even as y is so, m being prime to n one even and the other odd, hence we must have $pq=2mn$, and as q is odd this equation can subsist only thus, $q=ab$, $m=ac$, $n=bd$, and $\therefore$ $p=2cd$; a, b, c, d, being prime to each other and a and b odd, and either c or d even; the 2nd condition $m^2-n^2=z=2p^2+q^2$ then gives $a^2c^2-b^2d^2=8c^2d^2+a^2b^2$ or $\dfrac{c^2-b^2}{8c^2+b^2}=\dfrac{d^2}{a^2}$ and if there be a common measure, it must obviously be 3 or 9; if there be no CM we should have $c^2-b^2=d^2$ and $8c^2+b^2=a^2$ giving $\dfrac{a^2+d^2}{3}=3c^2$ which is impossible as a is prime to d, Art. 8, 1°. If 3 were CM we should have $c^2-b^2=3d^2$ and $8c^2+b^2=3a^2$ giving $a^2+d^2=3c^2$ which is impossible by the same Art. 8, 1°, or by Art. 14, 1°. If 9 were CM we should have $c^2-b^2=9d^2$ and $8c^2+b^2=9a^2$ giving $c^2-d^2=a^2$ and $c^2-9d^2=b^2$ two equations exactly similar to the proposed pair, and thereby proving as usual, that no solution can exist in great integers as none exists in small integers.

N.B.—As x, y, z, are prime to each other, and as ℣ 1st equation x divides y^2+z^2 $\therefore$ x cannot be divisible by 3 (Art. 8, 1° or Art. 14) $\therefore$ x is also prime to $3y$, and moreover as y was proved even, hence we could find the general solution of both the given equations by Art. 1, and then demonstrate this proportion otherwise.

39.—The two simultaneous equations $\left\{\begin{matrix} x^2-y^2=z^2 \\ x^2-10y^2=w^2 \end{matrix}\right\}$ are impossible.

Here $9y^2=z^2-w^2$, and $\therefore$ by Art. 5, if the proposed equations be possible, they will be so when x, y, z, and w, are all prime to each other, except perhaps, z and w, which if not prime to each other when the rest are so, could have no C M but 3 ($=\sqrt{9}$) and moreover x and w will then be prime to 10 (coeft. of $10y^2$) and $\therefore$ both odd, and $\therefore$ y even, ℘ 2nd equation, hence by Art. 1, the general solution of the 1st equation is $x=m^2+n^2, y=2mn$, and by Art. 2 the general solution of the 2nd equation is then $y=2pq$ and $x=10p^2+q^2$ or else $=2p^2+5q^2$, q being odd and prime to p, and m prime to n one even and the other odd, hence we must have $mn=pq$ which requires $m=ab$, $p=ac$ $q=bd$ and $n=cd$, a,b,c,d, being prime to each other and one of them (a or c) even, then the 2nd condition $m^2+n^2=x=10p^2+q^2$ gives $a^2b^2+c^2d^2=10a^2c^2+b^2d^2$, or $\therefore$ $\dfrac{d^2-a^2}{d^2-10a^2}=\dfrac{c^2}{b^2}$ and if there be a common measure, it must be 3 or 9; and if there be no C M we should have $d^2-a^2=\pm c^2$ and $d^2-10a^2=\pm b^2$, by taking the signs $+$, these two equations are similar to the original pair and then the usual argument proves the impossibility, but by taking the signs —, we get $b^2-c^2=\square$ and $b^2-10c^2=\square$ which are also similar to the original pair, so that then too the same argument still proves the impossibility; 3 could not be C M for by Art. 15 the equation $d^2-10a^2=\pm 3b^2$ is impossible; and if 9 were CM we should have $d^2-a^2=\pm 9c^2$ and $d^2-10a^2=\pm 9b^2$, these are similar to the proposed equations, when we take the sign $+$; and when we take the sign — they give $b^2-c^2=a^2$ and $b^2-10c^2=d^2$, which are also similar to the original pair.

But if we take $x=2p^2+5q^2$ then the 2nd condition becomes $a^2b^2+c^2d^2=2a^2c^2+b^2d^2$, or $\dfrac{a^2-5d^2}{2a^2-d^2}=\dfrac{c^2}{b^2}$ and here again, if there be a common measure, it could obviously be only 3 or 9; but this cannot be, for $2a^2-d^2$ is never divisible by 3 when a is prime to d; and if there be no C M we should have $a^2-5d^2=\pm c^2$ and $2a^2-d^2=\pm b^2$ giving $2c^2-b^2=\pm 9d^2$, which is impossible, for the very reason just now assigned, (Art. 8. 4°) and hence the two proposed equations are impossible simultaneously; and it may be observed that this demonstration leads very directly to the discovery of the *general theorem* in Art. 36; and since y was proved even, and $\therefore$ x,z, and w odd $\therefore$ the general solution of $9y^2=z^2-w^2$ can be got from Art 1, and thus we could easily demonstrate this theorem otherwise.

40.—The two simultaneous equations $\left\{\begin{matrix} x^2-y^2=\square=z^2 \\ x^2-12y^2=\square=w^2 \end{matrix}\right\}$ are impossible.

Here $11y^2=z^2-w^2$ and as no coeft. but 12 contains a square factor $\therefore$ by Art. 5, if the proposed equations be possible at all, they shall be so too, when x, y, z, w, are all prime to each other, except perhaps x and w, which, if not prime to each other when the rest are so, could then have no C M but 2, but even this exception cannot happen, as ℘ 1st equation x must be odd when prime to y and z, and x being odd the 2nd equation shows that w must be odd, and y even, and y being even the other three x,z,w, which are prime to y, must be odd, hence the general solution of the 1st equation is $x=m^2+n^2, y=2mn$ and by Art. 3, the general solution of the 2nd equation when y is even and prime to x and w, is $y=pq$ and $x=3p^2+q^2$, m being prime to n, and p to q, one of each pair being even and the other odd; hence we must have $pq=2mn$ and 1°—If p be odd, this equation can subsist only thus $p=ab$, $m=ac, n=bd$

and $\therefore q=2cd$ and then the 2nd condition $m^2+n^2=x=3p^2+q^2$ gives $a^2c^2+b^2d^2=3a^2b^2+4c^2d^2$, or $\therefore \frac{d^2-3a^2}{4d^2-a^2}=\frac{c^2}{b^2}$; a,b,c,d being prime to each other, and one of them (c or d) even; and if there be a C M it is plain it can only be 11; but this cannot be, for we should have $d^2-3a^2=\pm 11c^2$ and $4d^2-a^2=\pm 11b^2$, now taking the sign $+$, the equation $d^2-3a^2=11c^2$ or $\frac{d^2+c^2}{3}=4c^2+a^2$ is impossible by Art. 8, 1°. and when we take the sign $-$, we get $a^2+b^2=4c^2$ which is impossible when a is prime to b, as a^2+b^2 will then have the form $4n+1$ or $4n+2$ according as one or both of a and b are odd; and if there be no C M we should have $d^2-3a^2=\pm c^2$ and $4d^2-a^2=\pm b^2$, when we take the sign $+$ the latter of these equations is impossible for the reason just assigned, and when we take the sign $-$, the former equation $d^2-3a^2=-c^2$ or $d^2+c^2=3a^2$ is impossible by Art. 8, 1° or Art 14, 1°.—2° If q be odd, the equation $pq=2mn$ can subsist only thus, $q=ab, m=ac, n=bd$ and $\therefore p=2cd$ and then the 2nd condition gives $a^2c^2+b^2d^2=12c^2d^2+a^2b^2$ or $\frac{a^2-d^2}{a^2-12d^2}=\frac{c^2}{b^2}$; now if there be no C M we should have $a^2-d^2=\pm c^2$ and $a^2-12\,d^2=\pm b^2$, these are similar to proposed pair, when we take the sign $+$, and $\therefore$ by the usual argument the thing is impossible, and when we take the sign $-$, the 2nd giving $\frac{a^2+b^2}{3}=4d^2$ is impossible by Art. 8, 1° or Art. 14, 1°: if 11 be C M we should have $a^2-d^2=\pm 11c^2$ and $a^2-12d^2=\pm 11b^2$ giving $b^2-c^2=\pm d^2$ and $b^2-12c^2=\pm a^2$, which are similar to the equations of the foregoing case and $\therefore$ impossible for the reasons just assigned, and hence the proposed equations are impossible, Q. E. D.

41.—From the demonstration and final conditions of Articles, 35, 37, 38, we are naturally led to the discovery of the following.

GENERAL THEOREM.

The solution of $X^2-Y^2=Z^2$ and $X^2-(a+1)Y^2=W^2$ can be obtained from a solution of the *auxiliary* equations, $\left\{\begin{matrix} x^2-y^2=nz^2 \\ ax^2+y^2=nw^2 \end{matrix}\right\}$ or from a solution of the pair $\left\{\begin{matrix} y^2+x^2=nz^2 \\ y^2-ax^2=nw^2 \end{matrix}\right\}$ in fact I say $X=x^2w^2+y^2z^2$ and $Y=2xyzw$ will answer. Proof—It is self-evident that then $X^2-Y^2=(x^2w^2-y^2z^2)^2$ and so the 1st condition is actually fulfilled, and taking the 1st pair of auxiliary equations, $nX=x^2(ax^2+y^2)+y^2(x^2-y^2)=ax^4+2x^2y^2-y^4$ and $n^2Y^2=4x^2y^2(x^2-y^2)(ax^2+y^2)=4ax^6y^2-4x^4y^4(a-1)-4x^2y^6$ and $\therefore n^2(X^2-\overline{a+1}.Y^2)=a^2x^8+4ax^6y^2+(4-2a)x^4y^4-4x^2y^6+y^8-4a(a+1)x^6y^2+4(a^2-1)x^4y^4+4(a+1)x^2y^6=(ax^4-2ax^2y^2-y^4)^2$ and $\therefore X^2-(a+1)Y^2=(ax^4-2ax^2y^2-y^4)^2\div n^2 \therefore = \square$ and thus the 2nd condition is also fulfilled; and the demonstration will be much the same when we take the 2nd pair of auxiliary equations; see the general Theorem in Cor. 1, Art. 28.

Cor.—From the general Theorem given in Art. 36, in connexion with what has been demonstrated in the foregoing Articles, it follows that $x^2-ay^2=nz^2$ and $bx^2-y^2=nw^2$ are impossible, whatever rational numbers, a, b, and n may be, provided ab is $=$ any positive integer between 1 and 18, except 7 or 11;

and from the preceding general Theorem it follows also that $x^2-y^2=nz^2$ and $ax^2+y^2=nw^2$ are impossible; as also the pair $x^2+y^2=nz^2$, $x^2-ay^2=nw^2$ are impossible whatever rational number n may be when a is any positive integer < 12 except 6 or 10.

It is scarcely necessary to observe that by taking a or b negative in Art. 19, we could deduce the solution, $X^2+Y^2=Z^2$ and $X^2-abY^2=W^2$ from a solution of the *auxiliary* equations $x^2\mp ay^2=nz^2$ and $y^2\pm bx^2=nw^2$; so also by taking a negative in the Cor., Art. 19, we could deduce the solution $X^2+Y^2=Z^2$ and $X^2-aY^2=W^2$ from a solution of the auxiliary equations $x^2+y^2=nz^2$ and $x^2-ay^2=nw^2$. Again, by taking a negative in Art. 36, we could deduce the solution of $X^2-Y^2=Z^2$ and $X^2+abY^2=W^2$ from a solution of the two auxiliary equations $x^2+ay^2=nz^2$, $bx^2-y^2=nw^2$, and similar remarks will obviously apply to, and produce corresponding variations in the preceding general theorem, and to that given in Cor. 1, Art. 28.

We can find by the following method (analogous to that of Arts. 20 and 21) some integral values of A, so that the two equations $x^2-y^2=\square$ and $x^2-Ay^2=\square$ may be possible *simultaneously*; by taking $z=1$ in the two auxiliary equations $x^2-ay^2=nz^2$, and $bx^2-y^2=nw^2$ we get $n=x^2-ay^2$ and $\therefore b=w^2-\frac{y^2}{x^2}(aw^2-1)$ so that if $a=1$ then $A=ab=w^2-\frac{y^2}{x^2}(w^2-1)$ and here $w=3$ and $x=2$ gives $A=9-2y^2$; $w=5$ and $x=2$ gives $A=25-6y^2$; $w=7$ and $x=4$ give $A=49-3y^2$; $w=9$ and $x=4$ give $A=81-5y^2$; $w=8$ and $x=3$ give $A=64-7y^2$; $w=10$ and $x=3$ give $A=100-11y^2$; $a=2$, $w=5$, and $x=7$, give $b=25-y^2$ and $\therefore A=ab=50-2y^2$, &c., &c., so that the proposed equations $x^2-y^2=\square$ and $x^2-Ay^2=\square$ will be *possible* when A=7, 11, 18, 19, 22, 32, 36, 37, 42, 46, 48, 56, 57, 61, &c., &c. But here our observations shall terminate.

55, Grand Parade, Cork.
Nov. 20th, 1855.

APPENDIX.

The few properties of Numbers contained in the following short Appendix will form a useful sequel to this Tract, as by means of some of them several of the foregoing demonstrations may be much abbreviated; they may be also very acceptable and valuable to such readers of this Tract as have not an opportunity of consulting the elegant works of *Gauss*, *Barlow*, and *Legendre*, on the Theory of Numbers.

Article 1.—The product of *two* numbers of the form, a^2+nb^2 is itself of the same form: For $(a^2+nb^2)(c^2+nd^2)$ is obviously $=(ac\pm nbd)^2+n(ad\mp bc)^2$.

Cor.—Hence the product of *several* numbers, each of the form a^2+nb^2 is itself always of the same form, and that too in several different ways: and $\therefore$ (taking $n=1$) the product of several integers, each of which is the sum of two squares is also itself the sum of two squares in several ways.

Art. 2.—The product of two numbers, each of which is the sum of four integral squares, is itself also the sum of four integral squares, in several ways; For by page 28 of Brioschi's Treatise on Determinants,

$\begin{vmatrix} \alpha, & \alpha' \\ \beta, & \beta' \end{vmatrix} \times \begin{vmatrix} \gamma, & \gamma' \\ \delta, & \delta' \end{vmatrix} = \begin{vmatrix} \alpha\gamma+\alpha'\gamma', & \alpha\delta+\alpha'\delta' \\ \beta\gamma+\beta'\gamma', & \beta\delta+\beta'\delta' \end{vmatrix}$ that is $(\alpha\beta'-\alpha'\beta)\ (\gamma\delta'-\gamma'\delta = (\alpha\gamma+\alpha'\gamma')$ $(\beta\delta+\beta'\delta')-(\alpha\delta+\alpha'\delta')\ (\beta\gamma+\beta'\gamma')$ as is indeed evident by actual multiplication.

Now let $\alpha=a+b\sqrt{-1}$ also $\gamma=p+q\sqrt{-1}$
$\beta'=a-b\sqrt{-1}$ „ $\delta'=p-q\sqrt{-1}$
$\beta=c+d\sqrt{-1}$ „ $\delta=r+s\sqrt{-1}$
and $-\alpha'=c-d\sqrt{-1}$ and $-\gamma'=r-s\sqrt{-1}$

and the preceding equation becomes $(a^2+b^2+c^2+d^2)(p^2+q^2+r^2+s^2)=(ap-bq+cr-ds)^2+(aq+bp-cs-dr)^2+(ar-bs+pc+dq)^2+(as+br+cq+dp)^2$ which proves the theorem: and as the signs of a, b, c, d, p, q, r, s, may obviously be changed at random; $\therefore$ the preceding value of $(a^2+b^2+c^2+d^2)\ (p^2+q^2+r^2+s^2)$ can be expressed in various ways as the sum of four squares. By taking some or all of a, b, p, equal zero, it follows that the product obtained by multiplying the sum of two, three, or four integral squares, by another sum of two, three, or four squares, is always the sum of four or a less number of squares.

3.—Whenever $n=a^2+b^2$, x^2-ny^2 can be transformed into $nx'^2-y'^2$ and conversely, this last can be transformed into the former: For if x^2-ny^2 were $=nx'^2-y'^2$ we should have by transposition $x^2+y'^2=n(x'^2+y^2)$ i. e. $=(a^2+b^2)(x'^2+y^2)$ $\therefore$ $=(ax'-by)^2+(ay+bx')^2$ which is fulfilled if $x=ax'-by$ and $y'=ay+bx'$, and these give $x'=\dfrac{x+by}{a}$ and $y'=\dfrac{bx+ny}{a}$ or conversely $x=\dfrac{nx'-by'}{a}$ and $y=\dfrac{y'-bx'}{a}$ where of course a and b can be interchanged or change the signs of one or both; If $a=1$ these values of x' and y' in terms of x and y, et contra, will be *integral*. Ex. gr. x^2-2y^2 will be $=2x'^2-y'^2$ if $x'=x\pm y$ and $y'=2y\pm x$, or else $x=2x'\pm y'$ and $y=y'\pm x'$, and x^2-5y^2 will be $=5x'^2-y'^2$ if $x'=x\pm 2y$ and $y'=5y\pm 2x$, or else $x=5x'\pm 2y'$ and $y=y'\pm 2x'$, &c., &c.

4.—In Art. 14 of the foregoing Tract we proved that *every* divisor of x^2+y^2 must be of the same form when x is prime to y; and we shall now prove in a similar way, that every divisor of $x^2\pm 2y^2$ must itself be also of the same form when x is prime to y; for if A divides $x^2\pm 2y^2$ it must also divide $(x^2\pm 2y^2)$

$\times (a^2 \pm 2b^2) = (ax \pm 2by)^2 \pm 2(bx - ay)^2$ whatever integers a and b may be, but as x is prime to y, a and b can be found such that $bx - ay = \pm 1$, and thus A must divide $(ax \pm 2by)^2 \pm 2$, let q be the nearest integer to $\frac{ax \pm 2by}{A}$ whether $>$ or $<$, then $ax \pm 2by$ will $= Aq \pm p$ where $p < \frac{1}{2}A$, and as A divides $(ax \pm 2by)^2 \pm 2 = (Aq \pm p)^2 \pm 2 = A^2q^2 \pm 2Apq + p^2 \pm 2$ it must $\therefore$ divide $p^2 \pm 2$; now convert $\frac{p}{A}$ into a continued fraction and thence find the fractions converging to $\frac{p}{A}$; let $\frac{m}{n}$ and $\frac{m'}{n'}$ be two of these convergents, consecutive, and such that n is less, and $n' > \sqrt{A}$ which is always possible since 1 and A are the denominators of the first and last convergents $\left(\frac{0}{1} \text{ and } \frac{p}{A}\right)$; the difference of $\frac{p}{A}$ and $\frac{m}{n}$ is, by the property of the convergents, $< \frac{1}{nn'}$ and $\therefore \left(\frac{p}{A} - \frac{m}{n}\right)^2 < \frac{1}{n^2n'^2}$ or $(pn - Am)^2 < \frac{A^2}{n'^2}$ but by hyp. $n'^2 > A \therefore (pn - Am)^2 < A$ and as by hypothesis $n^2 < A$ Hence for the 1st Theorem $(x^2 + 2y^2)$ where we use the upper sign $(+)$ we have $(pn - Am)^2 + 2n^2 < 3A$. But the left hand member is divisible by A since as above $p^2 + 2$ is *then* divisible by A, we must $\therefore$ have $(pn - Am)^2 + 2n^2 = A$ or $2A$; if it be $= A$ the first theorem is proved, and if $(pn - Am^2) + 2n^2$ be $= 2A$, $pn - Am$ must be even, and then too the first theorem is proved, since A is then $= n^2 + 2\left(\frac{pn - Am}{2}\right)^2$. But for the 2nd theorem $(x^2 - 2y^2)$ where we use the lower sign $(-)$ we have $2n^2 - (pn - Am)^2 < 2A$, but the left-hand member is divisible by A since $p^2 - 2$ is *then* divisible by A, we must $\therefore$ have $2n^2 - (pn - Am)^2 = A$ which proves the 2nd theorem, since by Art. 3, $2n^2 - (pn - Am)^2$ is also of the form $x'^2 - 2y'^2$.

When p is $=$ or > 1 then $\frac{p}{A}$ will always give convergents, the first and last of which are $\frac{0}{1}$ and $\frac{p}{A}$, and $\frac{p}{A} - \frac{m}{n}$ or $\therefore pn - Am$ will not then be $= 0$ even if $\frac{m}{n}$ should be the first convergent $\frac{0}{1}$ and hence in the value of $A (= x'^2 \pm 2y'^2)$ neither x' nor y' can ever be $= 0$ unless when $p = 0$ and $\therefore A = 2$; (since A divides $p^2 \pm 2$ which is then $= 2$).

5. Every *odd* divisor of $x^2 + 3y$ or of $x^2 - 5y^2$ must be of the same form when x is prime to y: For if A divides $x^2 + 3y^2$ it must also divide $(x^2 + 3y^2)(a^2 + 3b^2) = (ax + 3by)^2 + 3(bx - ay)^2$ and taking a and b as before, such that $bx - ay = \pm 1$ and putting $ax + 3by = Aq \pm p$ where $p < \frac{1}{2}A$ we shall have $p^2 + 3$ divisible by A; then as before converting $\frac{p}{A}$ into a series of convergents (see Arts. 2 and 3 of *my Tract on Continued Fractions*) we find as above $(pn - Am)^2 < A$ and $n^2 < A$ and $\therefore (pn - Am)^2 + 3n^2 < 4A$, but the left hand member is divisible by A since $p^2 + 3$ is so divisible $\therefore (pn - Am)^2 + 3n^2$ must be $= A$, $2A$ or $3A$; if it be $= A$ the theorem is proved, and it could not be $= 3A$ unless $pn - Am$ were divisible by 3 and then $A = n^2 + 3\left(\frac{pn - Am}{3}\right)^2$ which

also proves the theorem in this case. Lastly, $(pn—Am)^2+3n^2$ could never be $=2A$ if A be *odd*, since $x'^2+3y'^2$ when even is never $=4r+2$ the double of an odd number (A). And as before, we may observe that in the value of the odd divisor A ($=x'^2+3y'^2$) neither x' nor y' will be $=0$ unless when $p=0$ and $\therefore$ $A=3$ (since A divides p^2+3 then $=3$); the demonstration is much the same for $x^2—5y^2$.

6.—Let Mp denote a multiple of the given integer p, and let X denote a polynomial of the mth degree $A_0x^m+A_1x^{m-1}+A_2x^{m-2}\ldots A_{m-1}x+A_m$ all whose coefts. are integers, then if $X=Mp$ be fulfilled when $x=$ the integer a it will also be plainly satisfied when $x=a\pm np$, n being *any wh. n°* which can be taken such that this value of x may lie between—$\frac{1}{2}p$ and $+\frac{1}{2}p$ or between 0 and p; and these integral values of x between 0 and p that fulfil the equation $X=Mp$ are called its *roots*; if all the coefts. $A_0, A_1, A_2\ldots A_m$ be divisible by p the proposed equation $X=Mp$ is obviously *identical* i. e. satisfied when x is *any* integer; and if all the coefts. but A_m be divisible by p the equation is evidently impossible; the roots of $X=Mp$ are evidently the same as those of $X+pX'=Mp$ et contra, X′ being any rational and integral function of x having all its coefts. integers; hence all the coefts. can be made $<p$ by increasing or decreasing them by proper multiples of p; when p is a prime number, X′ can be taken such that the first coeft. A_0 shall be $=1$: For by multiplying the given polynomial X by some integer $b<p$ and from the product deducting cpx^m the first coeft. A_0 is, in the result, changed into $bA_0—cp$ and as p is prime to A_0, b and c can be found such that $bA_0—cp$ shall $=\pm1$, and it is plain when $bX—cpx^m=Mp$ we must also have $X=Mp$.

Supposing then the first coeft. $A_0=1$ and the equation $X=Mp$ not *identical* I say it cannot then have more than m roots, x^m being the highest power of x in X; For if a be one root, let $\dfrac{X}{x—a}$ give the quotient X′ (of the m—1th degree) and the remainder R which does not contain x, then $X=(x—a)\,X'+R$ and by taking $x=a$, R must $=X_a$ viz.: the numerical value X takes when in it x is changed into a; thus $X=(x—a)\,X'+X_a$ and as this $=Mp$ when $x=a$ $\therefore$ $X_a=Mp$ and $\therefore$ $X=Mp$ requires that we should have $(x—a)X'=Mp$ so that if b be a 2nd root of $X=Mp$ we shall have $(b—a)\,X'_b=Mp$, and as $b—a$ is less than, and prime to p $\therefore$ we must have $X'_b=Mp$, i. e., b must be a root of the equation $X'=Mp$, x^{m-1} being the highest power of x in X′, and hence the equation $X=Mp$ of the mth degree can have only one root more than the equation $X'=Mp$ of the m—1th degree, and in like manner this last equation can have only one root more than an equation $(X''=Mp)$ of the degree $m-2$, &c., and as the final equation thus obtained $x—l=Mp$ can plainly have only one root (between 0 and p) $\therefore$ the proposed equation $X=Mp$ of the mth degree can never have more than m roots (between 0 and p), but it may have less than m roots, or even have no root.

Cor.—If the equation $X=Mp$, whose highest term is x^m have actually m roots $a, b, c, \ldots k, l$, these roots must manifestly be also roots of $X—(x—a)(x—b)\ldots(x—l)=Mp$ whose degree is only m—1, which by what was just proved, is impossible unless this last equation be *identical* and $\therefore$ X must $=(x—a)(x—b)\ldots(x—l)+p\,X'$, X′ being a rational and integral function of the degree m—1 having all its coefts. integers.—*FERMAT'S THEOREM.*

7.—If p be a prime number which does not divide the integer a it must divide $a^{p-1}—1$. Take the p—1 following multiples of a, viz., $a, 2a, 3a, \ldots (p—1)a$:

p cannot divide any of these numbers ma since p is prime to a and $> m$; neither can p divide the difference $ma—m'a$ of any pair of these multiples, since this difference is itself a term of the same series. If ∴ we take the least positive remainders of these multiples when divided by p, these remainders will be all different from each other, and as none of them can be zero ∴ they can only be the numbers 1, 2, 3 . . . $p—1$ arranged in a different order, and hence the remainder of $a \times 2a \times 3a \ldots (p-1)a \div p$ must be the same as the remainder of $1 \times 2 \times 3 \ldots (p-1) \div p$ and ∴ p must divide the difference $a \times 2a \times 3a \ldots (p-1)a - 1 \times 2 \times 3 \ldots (p-1) = 1.2.3 \ldots (p-1) \times (a^{p-1}-1)$ and as p is prime to $1 \times 2 \times 3 \ldots (p-1)$ ∴ it must divide $a^{p-1}-1$

Cor. 1—Hence the equation $x^{p-1}-1=Mp$ has $p—1$ roots 1, 2, 3, . . $p-1$ which are also obviously the roots of $x^{p-1}-1+pX=Mp$, X being any polynomial of the degree $p—1$ in x and having all its coefts. integers; now if X_m divides $x^{p-1}-1+pX$ giving X_n for quotient, m and n denoting the highest power of x in X_m and X_n respectively and ∴ $m+n=p-1$, then as $X_m X_n = x^{p-1}-1+pX$ ∴ the roots 1,2,3 . . . $p—1$ of $x^{p-1}-1+pX=Mp=X_m X_n$ can only be the roots of the 2 equations $X_m=Mp$ and $X_n=Mp$; and hence $X_m=Mp$ must have *exactly* m roots, and $X_n=Mp$ must have n roots *different* from the former; for by Art. 6, $X_m=Mp$ cannot have more than m roots, and if it had less than m roots, the other equation ($X_n=Mp$) should have more than n roots (which is impossible by Art. 6) as the equation $X_m X_n=Mp$ has altogether $m+n$ or $p—1$ roots ∴ &c.

Cor. 2.—Since the remainders of a, $2a$, $3a$ . . . $(p-1)a$ when divided by p are only the numbers 1,2,3, . . . $p—1$ arranged in a different order ∴ by raising each term of both series to any power q, p must divide the difference of the 2 sums, *i.e.* p must divide $(a^q-1)(1^q+2^q+3^q \ldots \overline{p-1}^q)$; Now if $q < p-1$ then, by Art 6, $x^q-1=Mp$ can only have q roots at most, by taking ∴ for a one of the numbers $<p$ which is *not* a root of $x^q-1=Mp$, a^q-1 will *not* be divisible by p and ∴ $1^q+2^q+3^q \ldots (p-1)^q$ *must* be divisible by p when $q<p-1$: If $q=m(p-1)$ then as by Art. 7, $a^{p-1}-1=Mp$ ∴ $a^{(p-1)m}=(1+Mp)^m$ ∴ $=1+Mp$ and now taking $a=1,2,3 \ldots p-1$ and adding the results, we find $1^q+2^q+3^q \ldots (p-1)^q=-1+Mp$ when $q=(p-1)m$; and as $a^{(p-1)m+n}=a^n+Mp$ ∴ $1^q+2^q+3^q \ldots (p-1)^q$ is always $=Mp$ whenever q is not divisible by $p—1$.

Cor. 3.—If $p>a$ be not a prime n°, but still prime to a, and, if instead of multiplying a by *all* the numbers $<p$ we only multiply it by the n numbers 1, a, b, c . . . $p—1$ which are less than p and *prime* to it, the reasoning of Art. 7 will still prove that p must divide a^n-1; a being *any* number prime to p; For when $a>p$ and $=mp+a'$ the remainder of $a^n \div p$ is manifestly the same as the remainder of $a'^n \div p$: this generalization of Fermat's famous theorem is due to Euler.

Cor. 4.—Since when p is a prime n°. the equation $x^{p-1}-1=Mp$ has the $p—1$ roots 1,2,3 . . . $p—1$ ∴ $(x-1)(x-2)(x-3) \ldots (x-p+1)-(x^{p-1}-1)=Mp$ which is only of the degree $p—2$ admits the same $p—1$ roots, and must ∴ be *identical* i.e. $=pX$ (by Cor. Art. 6) where X is a function of x, of $p—2$ dimensions, and having *integral* coefts; hence, then $(x-1)(x-2)(x-3) \ldots (x-p+1)=x^{p-1}-1+pX$ and ∴ if S_1 denote the sum of the numbers 1, 2, 3 . . $p—1$, S_2 = the sum of their products 2 by 2, S_3 = the sum of their products 3 by 3, and S_{p-1} = the continued product of them all;

also f_1 = their sum, f_2 = the sum of their squares, f_3 = the sum of their cubes &c., we shall have $S_1 = Mp$, $S_2 = Mp$, $S_3 = Mp$, &c., and $S_{p-1} + 1$ i.e. $1.2.3 \ldots (p-1) + 1 = Mp$: this last equation is Wilson's theorem; Newton's formulas $f_1 - S_1 = 0$; $f_2 - f_1 S_1 + 2S_2 = 0$; $f_3 - f_2 S_1 + f_1 S_2 - 3S_3 = 0$, &c., prove that $f_1, f_2, f_3, \ldots f_{p-2}$ must be also divisible by p since $S_1, S_2, S_3 \ldots S_{p-2}$ were just proved to be so divisible. See Cor. 2.

ANOTHER DEMONSTATION OF WILSON'S THEOREM.

Let a be *any* one of the numbers 1, 2, 3, . . . $p-1$, then by the proof of Art. 7, when the series $a, 2a, 3a \ldots (p-1)\,a$ is divided by p, some one of the remainders must be 1, let it be the remainder of Aa ∴ $Aa - 1 = Mp$ Now if $A = a$ we shall have $a^2 - 1 = (a+1)(a-1) = Mp$, and as p is a prime N°. it must ∴ divide $a+1$ or $a-1$; but as $p > a$ ∴ it cannot divide either of these factors unless $a-1=0$, or else $a+1=p$; hence, then A cannot $=a$ unless $a=1$ or $p-1$, thus the numbers 2, 3, 4, . . . $p-2$ can be associated in pairs so that the product of each conjugate pair$=1+Mp$ and ∴ the product of all these pairs i.e. $2 \times 3 \times 4 \ldots (p-2) = 1 + Mp$ which, multiplied by $p-1 = -1 + p$ gives $1.2.3.4. \ldots (p-1) = -1 + Mp$ or $1.2.3.4. \ldots (p-1) + 1 = Mp$. Q. E. D.

Obs.—This remarkable property beiongs to *prime numbers only;* for if θ be a divisor of p, then θ will obviously divide $1.2.3. \ldots (p-1)$ and ∴ θ cannot divide $1.2.3. \ldots (p-1)+1$, and ∴ a fortiori p cannot divide it.

Cor. 5.—Since $x^{p-1} - 1 = \left(x^{\frac{p-1}{2}} + 1\right)\left(x^{\frac{p-1}{2}} - 1\right)$ is divisible by the prime n°. p when x itself is not so divisible ∴ some one of the 2 factors $x^{\frac{p-1}{2}} \pm 1$ must then be divisible by p, and hence the remainder of $x^{p-1} \div p$ is always 0 or 1, and the remainder of $x^{\frac{p-1}{2}} \div p$ is always 0 or ± 1 i.e. 0, 1 or $p-1$; by taking $p=3, 5, 7, 11, 13, 17$, &c., we find

$x^2 \equiv 3n$ or $3n+1$, also $\equiv 5n$ or $5n \pm 1$	$x^8 \equiv 17n$ or $17n \pm 1$
$x^3 \equiv 7n$ or $7n \pm 1$	$x^9 \equiv 19n$ or $19n \pm 1$
$x^4 \equiv 5n$ or $5n+1$	$x^{10} \equiv 11n$ or $11n+1$
$x^5 \equiv 11n$ or $11n \pm 1$	$x^{11} \equiv 23n$ or $23n \pm 1$
$x^6 \equiv 7n$ or $7n+1$ also $\equiv 13n$ or $13n \pm 1$	$x^{12} \equiv 13n$ or $13n+1$

It follows, moreover, from Cor. 1, that each of the factors $x^{\frac{p-1}{2}} \pm 1$ has $\frac{p-1}{2}$ roots or values of x between 0 and p, that render them divisible by p, and in general, that $x^\theta - 1 = Mp$ has exactly θ roots when θ is a divisor of $p-1$; for $x^\theta - 1$ will then divide $x^{p-1} - 1$ since the difference of any 2 numbers (x^θ and 1) always divides the difference of the mth powers of those two numbers $(m = \frac{p-1}{\theta})$

Cor. 6.—Let x and y be two integers > 0 and $< p$, and θ a divisor of $p-1$, then if $x^\theta - y = Mp$ we must have $y^{\frac{p-1}{\theta}} - 1 = Mp$, and conversely if $y^{\frac{p-1}{\theta}} - 1 = Mp$, the equation $x^\theta - y = Mp$ will have θ roots.

1°. $x^\theta - y = Mp$ gives $(x\theta)^{\frac{p-1}{\theta}} = (y+Mp)^{\frac{p-1}{\theta}}$ i. e. $x^{p-1} = y^{\frac{p-1}{\theta}} + Mp$, but by Fermat's theorem $x^{p-1} = 1 + Mp$, and hence $y^{\frac{p-1}{\theta}} - 1$ must $= Mp$ which proves the 1st part of the theorem.

2°. $y^{\frac{p-1}{\theta}} - 1 = Mp$ say $= pQ$, shows that $x^{p-1} - y^{\frac{p-1}{\theta}}$ is the very same as $x^{p-1} - 1 - pQ$ and as $x^\theta - y$ divides $x^{p-1} - y^{\frac{p-1}{\theta}}$ (for the difference of two numbers always divides the difference of their *m*th powers) $\therefore$ $x^\theta - y$ also divides $x^{p-1} - 1 - pQ$ and $\therefore$ by Cor. 1, $x^\theta - y = Mp$ must have θ and no more roots, (see Art. 6).

Now since $x^{12} - 1 = (x^2)^6 - 1 = (x^3)^4 - 1 = (x^4)^3 - 1 = (x^6)^2 - 1 = M \times 13$ and $x^{18} - 1 = (x^2)^9 - 1 = (x^3)^6 - 1 = (x^6)^3 - 1 = (x^9)^2 - 1 = M \times 19$.

$\therefore x^2 \div 13$	must give	6 and no more different remrs. > 0	
$x^3 \div 13$	....	4	... ...
$x^4 \div 13$	...	3	... ...
$x^6 \div 13$	...	2 viz. $\pm$ 1, or 1 and 12.	
$x^2 \div 19$	..	9	... ..
$x^3 \div 19$	...	6	... ...
$x^6 \div 19$	..	3	... ...
$x^9 \div 19$	...	2, viz. $\pm$ 1, or 1 and 18; &c., &c.	

The remrs. of 4th and 6th powers to *any* modulus must of course be found among the remrs. of squares, since 4th and 6th powers are also squares; now since remr. of $x^2 \div p =$ remr. of $(p-x)^2 \div p$ or $=$ the remr. of $(mp \pm x)^2 \div p$ $\therefore$ to any modulus p whether prime or not the No. of different remrs. of $x^2 \div p$ cannot be more than $\frac{p+1}{2}$ but it may be less, such as when $p = 8, 12, 16$, &c.; moreover as $x + y$ $(= p)$ always divides $x^m + y^m$ when m is odd $\therefore$ if remr. of $x^m \div p$ be r the remr. of $y^m = (p-x)^m$ or of $(mp-x)^m$ when $\div$ p must be $-r$, whether the modulus p be prime or not, and thus the remrs. of *all odd powers*, x^{2n+1} divided by any modulus or divisor will always consist of conjugate pairs $\pm r$, thus *Ex. gr.* $x^3 \equiv 13n$, or $13n \pm 1$, or $13n \pm 5$; also $x^3 \equiv 19n$, or $19n \pm 1$, or $19n \pm 7$, or $19n \pm 8$; this property is analogous to that proved for squares in Cor. 2, Art. 14, of the foregoing Tract.

Cor. 7—By Fermat's Theorem $x^{p-1} - 1 = Mp$ when p is a prime number, and x not divisible by p, $\therefore$ $x\,(x^{p-1} - 1) = x^p - x$ must be divisible by p whether x be or be not divisible by p, and as $x^p - x$ is always even (and $\therefore$ divisible by 2) whether x be even or odd $\therefore$ $x^p - x$ is always divisible by $2p$; taking $p = 3, 5, 7$, &c., we find $x^3 - x \equiv 6n$, $x^5 - x \equiv 10n$, $x^7 - x \equiv 14n$, &c., the second equation $x^5 - x \equiv 10n$ proves that every 5th power is terminated with the same digit as its root; and as x has necessarily $2p$ forms to modulus $2p$ $\therefore$ x^p must also have $2p$ forms; we proved that $x^3 - x = (x-1)x(x+1)$ is always divisible by 6, but when x is odd $x^3 - x$ is plainly divisible by 8, since $x - 1$ and $x + 1$ are then both even, and as they differ by 2 $\therefore$ one of them must be divisible by 4, and the other by 2; hence when x is odd $x^3 - x$ must be divisible by 24, or $\therefore$ by any divisor of 24, for $\frac{P}{6} - \frac{P}{8} = \frac{P}{24}$

Cor. 8—By means of Fermat's Theorem, we immediately detect the impossibility of certain indeterminate equations of high degress; *Ex. gr.* $x^7-11=29y$ is impossible in whole numbers, since it gives $(x^7)^4=(11+29y)^4$ *i.e.* $x^{28}=11^4+29M$; now the proposed equation shows that x cannot be divisible by 29 and $\therefore$ by Fermat's theorem $x^{28}-1=M\times 29$ and $\therefore 11^4-1$ should be divisible by 29 if the proposed equation were possible, but this is not so $\therefore$ &c. Again $x^{15}-6=19y$ is impossible in whole numbers, for it gives $(x^{15})^6=(6+19y)^6 \therefore =6^6+19\times M$ but $(x^{15})^6=(x^{18})^5$ and as the proposed equation shows that x cannot be divisible by 19 $\therefore$ by Fermat's Theorem $(x^{18})^5=1+19m$, hence if the proposed equation were possible 6^6-1 should be divisible by 19, but this is not so $\therefore$ &c. Lastly, the equation $2x^6+3y^6=z^6$ is also impossible. For if this equation were possible at all, it should clearly be so, when x,y,z are whole numbers prime to each other $\therefore$ if x were divisible by 7, y could not be so, and then by Cor. 5, $2x^6+3y^6\equiv 7n+3$ which by Cor. 5, could never$=z^6$; and if y were divisible by 7 and x not so, $2x^6+3y^6$ would $\equiv 7n+2$ and $\therefore$ by the same Cor. 5 it could not then be$=z^6$; and if neither x nor y were divisible by 7, then $2x^6+3y^6$ would $\equiv 7n+5$, which by the same Cor. 5 could not even then$=z^6$ $\therefore$ &c.

Cor. 9—Every prime number p of the form $4n+1$ is the sum of 2 integral squares in one way only. For by Cor. 5, the equation $x^{\frac{p-1}{2}}+1=Mp=x^{2n}+1$ has $2n$ roots, let q be one of them, then p divides $q^{2n}+1=(q^n)^2+1^2$ and $\therefore$ by Art. 14 of the preceding Tract, p itself must be the sum of 2 integral squares; now if possible, let $p=a^2+b^2$ and also$=c^2+d^2$, then none of the 4, a,b,c,d, could $=0$ else p would be a square number and $\therefore$ not a prime number. Now as p $(=4n+1)$ is odd, a and b must be even and odd, and so must c and d; let a and c be even, b and d odd, then $a^2-c^2=d^2-b^2$ gives $\frac{a+c}{d+b}=\frac{d-b}{a-c}$ let each of these$=$the irreducible fraction $\frac{A}{B}$; $2f$ the G C M of the former and $2g$ the G C M of the latter $\therefore$ $\left.\begin{array}{l}a+c=2fA\\ a-c=2gB\end{array}\right|\begin{array}{l}d+b=2fB\\ d-b=2gA\end{array}\Big|$ these give $a=fA+gB$ and $b=fB-gA$ and $\therefore p=a^2+b^2=(f^2+g^2)(A^2+B^2)$ should have factors or divisors and could not $\therefore$ be a prime number if it were the sum of a pair of squares in more ways than one.

Cor. 10. Every prime number p of the form $8n+1$ is of the 3 forms a^2+b^2, a^2+2b^2 and a^2-2b^2. For by Cor. 5 the equation $x^{\frac{p-1}{2}}+1=Mp=x^{4n}+1$ has $4n$ roots >0 and $<p$, let q be one of them, then p divides $q^{4n}+1=(q^{2n})^2+1^2=(q^{2n}-1)^2+2(q^n)^2=(q^{2n}+1)^2-2(q^n)^2$ and as q^{2n} is prime to 1 and q^n is obviously prime to $q^{2n}\pm 1$ $\therefore$ by Art. 4 of this appendix p itself must be of each of the 3 forms a^2+b^2, a^2+2b^2 and a^2-2b^2.

Cor. 11. Every prime number p of the form $8n+3$ is also of the form a^2+2b^2. By Fermat's Theorem p divides $2^{p-1}-1=2^{8n+2}-1=\left(2^{4n+1}+1\right)\left(2^{4n+1}-1\right)$ but the factor $2^{4n+1}-1=2(2^{2n})^2-1^2$ being of the form $2a^2-b^2$ which is also of the form a^2-2b^2 (Art. 3) its divisors must be also of this form by Art. 4 and $\therefore$ when odd they must be of the form $8n\pm 1$ (since every odd square a^2

has the form $8n+1$). Hence then p, which is of the form $8n+3$ could not divide the factor $2^{4n+1}-1$ and $\therefore$ it must divide the other factor $2^{4n+1}+1=2(2^{2n})^2+1^2$ and $\therefore$ by Art. 4, p itself must be of the form a^2+2b^2

Any number of the form $8n+3$ could never be $\equiv a^2+b^2$ nor $\equiv a^2-2b^2$ since a^2+b^2 when odd $\equiv 8n+1$ or $8n+5$ and a^2-2b^2 when odd is $\equiv 8n\pm 1$, For every odd square a^2 has the form $8n+1$.

Any number of the form $8n+5$ could never $\equiv a^2\pm 2b^2$. For, as above, a^2-2b^2 when odd $\equiv 8n\pm 1$ and a^2+2b^2 when odd $\equiv 8n+1$ or $8n+3$.

Any number of the form $8n+7$ could never $\equiv a^2+b^2$ nor a^2+2b^2 for similar reasons.

Cor. 12.—Every prime number p of the form $8n+7$ is also of the form a^2-2b^2. For by Fermat's Theorem p divides $2^{p-1}-1=\left(2^{4n+3}+1\right)\left(2^{4n+3}-1\right)$ and $\therefore$ p must divide some one of these two factors. But p could not divide $2^{4n+3}+1 = 1+2\left(2^{2n+1}\right)^2$ since by Art. 4, every divisor of this quantity must be of the form a^2+2b^2 and this when odd is always of the form $8n+1$ or $8n+3$, hence then p which is of the for $8n+7$ could not divide $2^{4n+3}+1$ and $\therefore$ it must divide $2^{4n+3}-1$ and $\therefore$ it must also divide $2\left(2^{4n+3}-1\right)=\left(2^{2n+2}\right)^2-2\times 1^2$ and $\therefore$ by Art. 4, p itself must also be of the form a^2-2b^2.

But *any* number of the form a^2-2b^2 whether it be *prime* or *not* is also of this form in ways innumerable, since by Art. 1 $(a^2-2b^2)(x^2-2y^2)=(ax\pm 2by)^2-2(ay\pm bx)^2$ and as $x^2-2y^2=1$ is possible in an infinite number of ways (see Art. 14 of my *Tract on Continued Fractions*) $\therefore$ $a^2-2b^2=(ax\pm 2by)^2-2(ay\pm bx)^2$ is also possible in an infinite number of ways.

Hence also if p be a prime number $\equiv 8n\pm 1$, remainder of $2^{\frac{p-1}{2}}\div p$ will be $+1$; For then by Cors. 10 and 12 $p=a^2-2b^2$ and $\therefore$ $2b^2=a^2-p$ and raising each side to power $\frac{p-1}{2}$ gives $2^{\frac{p-1}{2}}b^{p-1}=a^{p-1}+\mathrm{M}(p)$ and as by Fermat's Theorem a^{p-1} and $b^{p-1}\equiv 1+\mathrm{M}(p)$ $\therefore$ $2^{\frac{p-1}{2}}$ must also obviously $=1+\mathrm{M}(p)$. But if $p\equiv 8n+3$, then by Cor. 11, $p=a^2+2b^2$ $\therefore$ $2b^2=p-a^2$ and $\therefore$ $2^{\frac{p-1}{2}}\times b^{p-1}=\mathrm{M}(p)-a^{p-1}$ since $\frac{p-1}{2}$ is then odd; and as before a^{p-1} and $b^{p-1}\equiv 1+\mathrm{M}(p)$ by Fermat's Theorem it follows $\therefore$ that $2^{\frac{p-1}{2}}+1=\mathrm{M}p$ i.e. when $p=8n+3$ then remainder of $2^{\frac{p-1}{2}}\div p$ is -1 or which is the same, $p-1$, and the same is true if $p=8n+5$.

Cor. 13.—Every prime number p of the form $3n+1$ is also of the form a^2+3b^2.

Every *prime* number >3 is obviously of the form $6n+1$ or else $6n-1$ and $\therefore$ as p is of the form $3n+1$ we must have $p=6n+1$; now $x^{p-1}-1=x^{6n}-1=(x^{2n}-1)(x^{4n}+x^{2n}+1)$ $\therefore$ by Cor. 1, $x^{4n}+x^{2n}+1=\mathrm{M}p$ has $4n$ roots >0 and

$< p$, let q be one of them $\therefore p$ divides $q^{4n}+q^{2n}+1=(q^{2n}-1)^2+3(q^n)^2$ and as $p=6n+1$ is odd $\therefore$ by Art. 5 of this Appendix, p itself must be of the form a^2+3b^2.

Cor. 14: $+2$ is remainder of $x^2\div$ any number N which is not divisible by 4, nor by any number of the form $8n+3$ or $8n+5$, and it is not a quadratic remainder of any other numbers. For if N be odd its prime factors can then be only of the form $8n+1$ or $8n+7$, each of which, by Cor. 10 and 12 is of the form a^2-2b^2; and, $\therefore$ by Art. 1, their product N must $=A^2-2B^2$, A being odd since N is so, $\therefore$ N divides $(A^2-2B^2)(x^2-2y^2)=(Ax-2By)^2-2(Bx-Ay)^2$, and if N contains no square factor A must be prime to B, and then x and y can be found such that $Bx-Ay=\pm1$, and thus N will divide $(Ax-2By)^2-2$ i.e. 2 will be the remainder of $(Ax-2By)^2\div N$; if N be even its only additional prime factor is 2 since by *hyp*, N is not divisible by 4, and $\therefore$ $N=(A^2-2B^2)2=2A^2-(2B)^2$ which, by Art. 3, is still of the form A^2-2B^2, where A must now be even and B odd, as N is divisible by 2, but not by 4; thus, then N divides $(A^2-2B^2)(x^2-2y^2)$, and, $\therefore$ as before, 2 must be remainder of $(Ax-2By)^2\div N$. Conversely, if N divides x^2-2, then every divisor of N will also divide x^2-2, and, $\therefore$ by Art. 4, N and every divisor of N $\equiv a^2-2b^2$, and if N be odd a must be so, and then N or any divisor of it $\equiv 8n\pm1$; but if N be even x must be so, and $=2x'$, then N divides $2(2x'^2-1)$ $\therefore \frac{1}{2}$N must be odd, and divide $2x'^2-1$, and be $\therefore$ itself and all its divisors of the form $a^2-2b^2\equiv 8n\pm1$.

Cor. 15: -2 cannot be remainder of $x^2\div N$ if N be divisible by 4, or by any factor $\equiv 8n+5$ or $8n+7$, and -2 is remainder of $x^2\div N$ when N is not divisible by 4 nor by any factor $\equiv 8n+5$ or $8n+7$. For if -2 were remainder of $x^2\div N$ then x^2+2 should be divisible by N, and by every divisor of N; but x^2+2 is never divisible by 4, since, when even, its half is odd; and as every divisor of x^2+2 is of the form a^2+2b^2 (by Art. 4), and this when odd $\equiv 8n+1$ or $8n+3$ which demonstrates the 1st part of the theorem. Conversely, if N be not divisible by 4, nor by any factor of the form $8n+5$ or $8n+7$, then the *prime* factors of N (except 2) being of the form $8n+1$ or $8n+3$, each of which $\equiv a^2+2b^2$ (by Cor 10 and 11) $\therefore$ by Art. 1, their product $N=A^2+2B^2$, and if, moreover, 2 be a factor, then $N=(2B)^2+2A^2$. Hence, N must divide $(A^2+2B^2)(x^2+2y^2)=(Ax+2By)^2+2(Ay-Bx)^2$; and if N contains no square factor A must be prime to B and then x and y can be found so that $Ay-Bx=\pm1$, and thus -2 will manifestly be the remainder of $(Ax+2By)^2\div N$.

Cor. 16: -3 cannot be remainder of $x^2\div N$ if N be divisible by 8 or 9, or by any factor of the form $6n+5$, and -3 will always be remainder of $x^2\div N$ whenever N is not divisible by 8, or 9, or $6n+5$. For if -3 be such a remainder, then x^2+3 should be divisible by N and by every diviser of N; but x^2+3 is never divisible by 8, for if it were x should be odd; but then $x^2+3\equiv 8n+4$ which is not divisible by 8. Neither is $x^2\pm3$ ever divisible by 9, since to be so x should $=3x'$, but even then $x^2\pm3=9x'^2\pm3$ is not divisible by 9, and as every *odd* divisor of $x^2+3=x^2+3\times1^2$ is of the form a^2+3b^2 (by Art. 5), if a be divisible by 3, b cannot be so since x^2+3 is never divisible by 9, then $a^2+3b^2=3n'=6n$ or $6n+3$; but if b be divisible by 3, a cannot be so, and then by Cor. 4, $a^2+3b^2=3n'+1=6n+1$, or $6n+4$; hence, a^2+3b^2 when odd and not divisible by 9 is of the form $6n+1$ or $6n+3$, and never of the form $6n+5$, so that no divisor of x^2+3 can ever be of the form $6n+5$ and $\therefore$ N cannot divide x^2+3 if N be divisible by 8, 9 or $6n+5$. Conversely, when N is not divisible by 8, 9 or any number of the form $6n+5$, then its prime factors can be only 2, 3, 4, and primes of the form $6n+1$, and as, by Cor. 13, *every* prime number of this form $\equiv a^2+3b^2$ $\therefore$ the product of all the factors of the forn $6n+1$ will $=A^2+3B^2$, which, multiplied by 3, is still of the same form A^2+3B^2, and as this must be odd and not divisible by 9 $\therefore$ A must be prime to B, one even and the other odd, and then $N= 4(A^2+3B^2)=(1^2+3\times 1^2)(A^2+3B^2)=(A+3B)^2+3(A-B)^2=P^2+3Q^2$, where P must be prime to Q, since A is prime to B, one even and the other odd; hence then N divides $(P^2+3Q^2)(x^2+3y^2)=(Px+3Qy)^2+3(Py-Qx)^2$, and as P is prime to Q $\therefore$ x and y can be found so that $Py-Qx=\pm 1$, and then -3 is plainly the remainder of $(Px+3Qy)^2\div N$. If N be divsible by 2, but not by 4, then 2N will divide $(Px+3Qy)^2+3$, and $\therefore$ N must divide it also, $\therefore$ &c.

Cor. 17: $+3$ is remr. of $x^2\div N$ whenever N is not divisible by 4 or 9, or by any odd n°. $\equiv 12n\pm 5$, and it is not remr. of $x^2\div N$ whenever N possesses any of these disqualifications.

For if $+3$ be remr. of $x^2\div N$ then x^2-3 should be divisible by N or $\therefore$ by any divisor of N; now x^2-3 is never divisible by 4, for if it were x should be odd, but then $x^2-3\equiv 8n-2$ which is not divisible by 4; and in the foregoing Cor. x^2-3 was proved to be never divisible 9; and as every divisor N (or else 2N) of $x^2-3\equiv 3a^2-b^2$ $\therefore$ every *odd* divisor of x^2-3 is either $=3a^2-b^2$ or $\frac{1}{2}(3a^2-b^2)$. Now to have $3a^2-b^2$ odd, a and b must be one even and the other odd; if a be even and b odd, and $\therefore$ $b^2\equiv 24n+1$ or $24n+9$ then $3a^2-b^2\equiv 12n+3$ or $12n+11$, but if a be odd and b even, then as every even square $=12n$ or $12n+4$ $\therefore$ $3a^2-b^2$ is then also $\equiv 12n+3$ or $12n+11$, and so when $3a^2-b^2$ is odd it is always $\equiv 12n+3$ or $12n+11$; and if $\frac{1}{2}(3a^2-b^2)$ be odd then a and b cannot be both even, nor one even and the other odd, they must $\therefore$ be both odd, and so a^2 or $b^2\equiv 24n+1$ or $24n+9$ and $\therefore$ $3a^2-b^2\equiv 24n+2$ or $24n+18$ and so when $\frac{1}{2}(3a^2-b^2)$ is odd, it is $\equiv 12n+1$ or $12n+9$ and we already proved that $3a^2-b^2$ when odd $\equiv 12n+3$ or $12n+11$. Henee then no odd divisor of x^2-3 is ever $\equiv 12n\pm 5$ and $\therefore$ $+3$ could never be remr. of $x^2\div N$ if N be divisible by 4 or 9 or by any number of the form $12n\pm 5$.

Cor. 18.—If a prime number $p=2n+1$ be of the form $8m+3$ or $8m+5$ then 2^n+1 will be divisible by p, otherwise 2^n-1 will be divisible by p. By Cor. 14, 2 cannot be remr. of $y^2\div p$ when $p=8m\pm 3$ and $\therefore$ by Cor. 6 the non-quadratic remr. 2 must be root of $x^n+1=Mp$ (since the n remrs. of $y^2\div p$ *are* roots of $x^n-1=Mp$) $\therefore$ p then divides 2^n+1 in like manner by Cor. 6, 10, 12 and 14, p must divide 2^n-1 when p or $2n+1$ is of the form $8m\pm 1$.

Cor. 19.—A prime n°. $p=2n+1$ of the form $12m\pm 5$ will always divide 3^n+1 and if p be not of this form it will divide 3^n-1.

By Cor. 6 the roots of $x^n-1=Mp$ are the n remainders of $y^2\div p$ and $\therefore$ the roots of $x^n+1=Mp$ are the remaining n numbers $< p$ which are not remainders of $y^2\div p$; now when $p=12m\pm 5$, it is proved in Cor. 17 that $+3$ cannot be remainder of $y^2\div p$ and $\therefore$ p *must then* divide 3^n+1: in like manner when p is not $=12m\pm 5$ it must divide 3^n-1.

Cor. 20.—If a whole n°. $p>2$ divides x^n-1, n being odd then p cannot divide x^m+1. For the equation $x^n-1=Mp$ gives $x^n=1+Mp$ and $\therefore$ $x^{mn}=(1+Mp)^m$ $\therefore$ $=1+Mp$; now if possible let $x^m+1=Mp$, transpose $+1$ and raise both sides to the *odd power n* and we find $x^{mn}=(-1+Mp)^n=-1+Mp$, but this cannot be since we already proved $x^{mn}=1+Mp$ $\therefore$ &c. *Ex. gr.*

$10^{15}-1 \div 31, 10^{3}-1 \div 37, 10^{5}-1 \div 41, 10^{21}-1 \div 43, 10^{13}-1 \div 53, 10^{33}-1 \div 67, 10^{35}-1 \div 71, 10^{13}-1 \div 79$ and $10^{41}-1 \div 83$ are all integers and as the powers of 10 are all odd numbers in these instances $\therefore$ $10^{n}+1$ can never be divisible by 31, 37, 41, 43, 53, 67, 71, 79 or 83, and $\therefore$ in general any whole number whose 1st and last figures are alike, and all whose middle figure are ciphers (0^s) can be never divisible by any of the integers above-mentioned.

Cor. 21.—Every prime number p which divides $a^{n}+1$ must be of the form $2nx+1$ or else it must also divide $a^{\omega}+1$ where $\frac{n}{\omega}=$ an odd number. For let $p=2nx+\pi$ and as this divides $a^{n}+1$ $\therefore a^{n}=-1+Mp$ and $\therefore a^{2nx}=1+Mp$; but by Fermat's theorem $a^{2nx+\pi-1}=1+Mp$, i. e. $a^{\pi-1}\times(1+Mp)=1+Mp$ and $\therefore a^{\pi-1}=1+Mp$ and this is fulfilled if $\pi=1$ or $\therefore p=2nx+1$.

But if $\pi>1$ let $\omega=$GCM of n and $\pi-1$ so that $n=n'\omega$ and $\pi-1=\pi'\omega$ which gives $a^{n'\omega}=-1+Mp$ and $a^{\pi'\omega}=1+Mp$; and now as n' is prime to π' we can have $fn'=g\pi'+1$ and $\therefore a^{fn'\omega}=a^{g\pi'\omega+\omega}=a^{\omega}\times a^{g\pi'\omega}$ *i. e.* $(-1+Mp)^{f}=a\omega(1+Mp)^{g}$ which gives $a^{\omega}=(-1)^{f}+Mp$ and thence $a^{n}=a^{n'\omega}$ $\therefore =(-1)^{n'f}+Mp$ but a^{n} is given $=-1+Mp$. Hence then the index $n'f$ must be odd, and $\therefore n'=\frac{n}{\omega}$ must also be odd. Q.E.D.

Obs.—Hence if n be a prime number, every divisor of $a^{n}+1$ must divide $a+1$ or else be of the form $2nx+1$: if $n=2^{m}$ then $a^{n}+1$ can have as divisors only prime numbers $\equiv 2nx+1$ as n is not then divisible by any *odd* number. If $n=2^{m}n'$, n' being a prime number, then a prime divisor of $a^{n}+1$ will $\equiv 2nx+1$ or else it must divide $a^{2^{m}}+1$ and must $\therefore \equiv(2^{m})2x+1$. If $n=m'n'$, m' and n' being odd primes, the prime divisor of $a^{n}+1$ will $\equiv 2nx+1$ or else it will divide $a^{m'}+1$ and $\therefore$ be $\equiv 2m'x+1$, or else it must divide $a^{n'}+1$ and $\therefore$ be $\equiv 2n'x+1$ or else it must divide $a+1$.

If $a=2$ and n be odd the divisors of $a^{n}+1$ will also divide $(a^{n}+1)a=x^{2}+2$ and can $\therefore$ be only of the form $a^{2}+2b^{2}\equiv 8m+1$ or $8m+3$ and never $\equiv 8m+5$ or $8m+7$.

Cor. 22.—Every prime number p which divides $a^{n}-1$ must $=nx+1$ or else it must divide $a^{\omega}-1$ where ω is a submultiple of n; if n be odd then the form $nx+1$ becomes $2nx+1$ and then too p must divide $x^{2}-a$. Proof.—Put $p=nx+\pi$ where $\pi<n$ $\therefore a^{n}=1+Mp$; and by Fermat's theorem $a^{p-1}=1+Mp$ but $a^{nx+\pi-1}=a^{\pi-1}\times a^{nx}$ and $\therefore a^{\pi-1}(1+Mp)^{x}=1+Mp$ and $\therefore a^{\pi-1}=1+Mp$ which is fulfilled if $\pi=1$ and $\therefore p=nx+1$: now when n is odd x must be even or else p would be so, which cannot be as p is prime $\therefore$ when n is odd $p=2nx+1$.

If $\pi>1$ let ω be G C M of n and $\pi-1$, and find f and g such that $fn-g(\pi-1)=\omega$; then the 2 equations $a^{n}=1+Mp$ and $a^{\pi-1}=1+Mp$ give $1+Mp=a^{fn}=a^{g(\pi-1)+\omega}=a^{\omega}(1+Mp)$ $\therefore a^{\omega}=1+Mp$, i. e. p will divide $a^{\omega}-1$ and as the same theorem applies to $a^{\omega}-1$ $\therefore$

1°. If n be prime, all the divisors of $a^{n}-1$ are comprised in $2nx+1$ or else they must divide $a-1$.

2°. If $n=m'n'$, m' and n' being prime numbers >2 then the divisor p of $a^{n}-1$ will $\equiv 2nx+1$ or else it will divide $a^{m'}-1$ and be $\therefore \equiv 2m'x+1$ or else it will divide $a^{n'}-1$ and be $\therefore \equiv 2n'x+1$ or else it must divide $a-1$.

3°. If $n=2^m$ the prime divisor p of a^n-1 will $\equiv nx+1$ or else it will divide $a^{\frac{1}{2}n}-1$ and be $\therefore\ \equiv\frac{1}{2}nx+1$ or else it will divide $a^{\frac{1}{2}n}+1$ and be $\therefore$ $=\frac{1}{4}nx+1$, &c., &c.

NOTE.—In Art. 21, page 20, when we took $z=1$ in the first auxiliary equation, the 2nd auxiliary equation gave $b=w^2+\frac{y^2}{x^2}(aw^2-1)$, and if we take $a=-3$, $w=15$ and $x=26$, this gives $b=15^2-y^2$ and $\therefore$ $a'=ab=3(y^2-15^2)$ or 3 (A^2-15^2) which when $A=16$ gives $a'=93$ a number which escaped the profound researches of the *celebrated Euler*, and which, together with the numbers 23 and 77 were communicated to me (*without demonstration*), by Charles H. Brookes, Esq., of Newcastle-upon-Tyne; if $a=-6$, $w=2$, and $x=5$ the same formula gives $b=4-y^2$ and $\therefore$ $a'=ab=6A^2-24$; again $a=-7$, $w=3$, and $x=8$ gives $b=9-y^2$ and $a'=ab=7(A^2-9)$; also $a=-8$, $w=1$, $x=3$ gives $a'=8(A^2-1)$ where $A(=y)$ must *not* be taken $=3$; again $a=-11$, $w=3$ and $x=10$ gives $b=9-y^2$ and $\therefore$ $a'=ab=11(A^2-9)$ $\therefore$ $=77$ if $A=4$, &c., &c.

Euler's list of the integral values of a that render the equations $x^2+y^2=\square$ and $x^2+ay^2=\square$ *simultaneously* possible is both *erroneous and defective*, as it gives 64 but not 93 for a.

We shall now prove a theorem which was extremely useful to us in the foregoing Tract, and whose demonstration was inadvertently omitted; the theorem is that alluded to in Art. 15, and given in Arts. 52 and 53 of *Barlow's* Theory of Numbers. Let A (>0) be one of the remrs. of x^2 divided by a prime number p, as modulus or divisor, and let B be one of the other $\frac{p-1}{2}$ integers between 0 and p which are not remrs. of $x^2\div p$, then I say AB will be non-residue of $x^2\div p$. Since A *is* a quadratic residue $\therefore$ $A=x^2-Mp$, and if possible let AB be also a quadratic residue and $\therefore$ $=y^2-Mp$; now as p is *prime*, and does not divide x (since A is *not*$=0$) $\therefore$ we could find a and b such that $y=ax-bp$ which gives $y^2=a^2x^2-Mp$, i.e $=a^2(A+Mp)-Mp$ $\therefore$ $=a^2A+Mp$ and then $AB=y^2-Mp$ $\therefore$ $=a^2A-Mp$ gives $A(a^2-B)=Mp$ and as p is prime to A $\therefore$ it should divide a^2-B, and so B should be remr. of $a^2\div p$ which is contrary to hypothesis $\therefore$ &c., hence then $(px+A)(py+B)\equiv pz+B$. The same theorem is also easily proved *directly* in page 72 of *Gauss's Recherches Arithmetiques.*

THE END.

CONTENTS OF THIS TRACT.

The foregoing little Tract is divided into three chapters.

Chap. I. treats of the possible and impossible cases of the two simultaneous equations $x^2+ay^2=\square$ *and* $x^2-ay^2=\square$ *; strict demonstrations of the impossibility by* one uniform method *are given when a is any whole number* <20 *except* 5, 6, 7, 13, 14, or 15 *; and in these cases that are possible general formulæ are given in Art.* 10 *for finding* immediately, without the tedious round-about of Fermat's method, *as many answers as we please in integers prime to each other ; and in Art.* 16 *it is proved that the two equations will be always impossible simultaneously, whenever a is a prime number, such that neither* m^2+1 *nor* m^2-2 *is divisible by a, m being* $<\frac{1}{2}a$.

Chap. II. begins at Art. 17 *and treats of the possible and impossible cases of the two simultaneous equations* $x^2+y^2=\square$ *and* $x^2+ay^2=\square$ *; rigorous demonstrations of the impossibility are given,* by one uniform method, *when a is any integer between* 1 *and* 20 *except* 7, 10, 11, *or* 17 *; and for these possible cases, general formulæ are given in Art.* 19 *for finding immediately, by a method* far more direct and expeditious than Fermat's *as many answers as we please in integers prime to each other ; in Cor.* 3, *Art.* 18, *it is proved that "there cannot be four square numbers in arithmetical progression," and in Arts.* 20 *and* 21 *it is proved that there are* 39 *positive integral values of a between* 1 *and* 100 *which render the proposed equations simultaneously possible ;* Euler's *list of these* 39 *integers proved* erroneous and defective. *To this chapter is subjoined a scholium containing very extensive and important additions to the* few *cases in which the equation* $ax^4+bx^2y^2+cy^4=\square$ *was heretofore known and proved* (by Fermat and Euler) *to be impossible.*

Chap. III. contains a similar minute and ample discussion of the possible and impossible cases of the two simultaneous $x^2-y^2=\square$ *, and* $x^2-ay^2=\square$ *. in which it is* rigorously demonstrated, by one uniform method, *that these two equations are impossible simultaneously when a is any integer between* 1 *and* 18 *except* 7 *or* 11.

CONTENTS OF THE APPENDIX.

Art. 1.—*The product of two or more numbers of the form* a^2+nb^2 *is itself of the same form.*

2.—*The product of two numbers, each of which is the sum of four integral squares is also itself the sum of four integral squares in various ways.*

3.—*Whenever* $n=a^2+b^2$, x^2-ny^2 *can be transformed into* $nx'^2-y'^2$ *et contra.*

4.—*Every divisor of* $x^2\pm2y^2$ *must be itself of the same form when x is prime to y.*

5.—*Every* odd *divisor of* x^2+3y^2 *or of* x^2-5y^2 *must be itself of the same form when x is prime to y.*

6.—*The equation* $X=Mp$ *cannot have more than m roots or values of x between* 0 *and p ;* Mp *denoting a multiple of a* prime *number p, and* X *being a function of x of the mth degree, having integral coefts.*

7.—*Easy proof of Fermat's famous theorem, to which are added* 22 *corollaries containing very concise and easy demonstrations of some of the most useful and remarkable propositions in the theory of numbers.*

Note on Art. 21, *containing also an easy proof of the useful theorem alluded to in Art.* 15.

ABBREVIATIONS.

The sign $(\div)$ of division is often used in the foregoing pages for "*is divisible by.*"

Gauss' sign $(\equiv)$ when used here signifies "*of the form.*"

In the Appendix M*p* or M(*p*) denotes "*a Multiple of p.*"

GCM or CM denote "*greatest common measure,*" or "*common measure.*"

In pages 54 and 55, CF denotes "*Continued Fraction.*"

CONTENTS OF MR. COLLINS' TRACT ON CONTINUED FRACTIONS.

Art. 1—An expression of the form $(x=)a_1 + \cfrac{1}{a_2 + \cfrac{1}{a_3 + \cfrac{1}{a_4 \&c.}}}$ where a_1, a_2, a_3, &c., are positive integers is called a *continued fraction ;* the use of continued fractions very natural in approximating to the value of any quantity x whether rational or not.

2—How to convert a Vulgar fraction into a Continued fraction.

3—If $\frac{p_1}{q_1}=a_1, \frac{p_2}{q_2}=a_1+\frac{1}{a_2}, \frac{p_3}{q_3}=a_1+\cfrac{1}{a_2+\cfrac{1}{a_3}}$ &c., then $\frac{p_1}{q_1}, \frac{p_2}{q_2}, \frac{p_3}{q_3}$, &c., are called converging fractions or convergents to the value of x; easy proof of the rule for calculating the convergents from a_1, a_2, a_3 &c., being given; the convergents are alternately $<$ and $>x$.

4—Easy proof of the remarkable equation $p_n q_{n-1} - p_{n-1} q_n = (-1)^n$; the converging fractions are ∴ irreducible; proof of the curious equation $\frac{p_n}{q_n} = a_1 + \frac{1}{q_1 q_2} - \frac{1}{q_2 q_3} + \frac{1}{q_3 q_4} \ . \ . \ . \ \pm \frac{1}{q_n q_{n-1}}$; a convergent $\frac{p_n}{q_n}$ which immediately precedes a large denominator a_{n+1} must be a very close approximation to (x) the value of the entire continued fraction.

5—Use of continued fractions in finding the integral solutions of indeterminate equations of the 1st degree, having 2 unknown quantities; the number of possible integral solutions of $ax+by=c$ is the the nearest integer to $\frac{c}{ab}$ when a is prime to b.

6—Each convergent $\frac{p_n}{q_n}$ approaches nearer to the true value of the sought quantity x than can be done by any other fraction $\frac{P}{Q} >$ or $<x$ provided $Q<q_n$ the same is true even when Q is $>q_n$ but $<q_{n+1}$ provided $\frac{p_n}{q_n}$ and $\frac{P}{Q}$ are both $>$ or else both $<x$; error of *Lagrange* about this matter.

7—Examples of the great use of convergents on account of the important property mentioned in Art. 6. Ex 1, finds the convergents $\frac{22}{7}, \frac{355}{113}$ &c., to the value of $\pi=3.1415926535898$, &c.

Ex. 2. finds the convergents $\frac{1}{4}, \frac{8}{33}$ &c., to .24224 Assuming the length of the Tropical year to be 365.24224 days. Hence is derived the Julian and Persian correction of the calendar; small error of the present Gregorian year.

Ex. 3 finds the convergents $\frac{99}{8}, \frac{235}{19}$ &c., to $365.25 \div 29.5305886 =$ ratio of the Julian year to a Synodic month. Hence the Octæteris of the ancient Greeks and the Metonic Cycle of 19 years; at the end of every 19 years the new moon falls on the same days of the months; at the end of 308 years the new moon falls one day earlier than is indicated by the preceding rule of Meton proemptose and Metemptose explained; at the end of every 16 Julian years, the new moon falls 3 days later in the months, great use of this fact in placing the Golden Numbers opposite the days of the months in the *Church Calendar*.

8.—The successive convergents approach more and more nearly to the true value of x which always lies between any consecutive pair of them; error of $\frac{p_n}{q_n}$ is $< \frac{1}{a_{n+1} q_n^2}$ Hence the reason why continued Fractions are applicable to the solution of indeterminate equations of the 2nd degree and inapplicable for higher degrees; amount of $Ap^2+Bpq+Cq^2$ is $< \frac{\sqrt{B^2-4AC}}{a_n}$ when $\frac{p}{q}$ is a convergent to either root of $Ax^2+Bx+C=o$, $\frac{p}{q}$ being followed by the next denominator a_n of the C.F for x; no denominator a_n of the C F for x can ever exceed $\sqrt{B^2-4AC}$; p^2-Nq^2 always $<2\sqrt{N}$, $\frac{p}{q}$ being a convergent to $\sqrt{N}$ converse theorem of *Lagrange* most useful and important.

9.—Min^m. value of p^3-Aq^3 when p, q, and A are integers, and A not a cube number, obtained by taking $\frac{p}{q}$ a convergent to $\sqrt[3]{A}$; amount of this min^m. $<3A^{\frac{1}{3}} \times \frac{p}{a}$ or $3A^{\frac{2}{3}} \times \frac{q}{a}$; a being the denominator consecutive to $\frac{p}{q}$ of the CF for $A^{\frac{1}{3}}$; this min^m. amount increases indefinitely with p; best numerical approximations to the duplication of the cube; $635^3 < 2 \times 504^3$ by less than the millionth part of the greater cube (635^3) the equation $p^n - Aq^n = B$; A and B being given integers and B small, can have no more than one or two solutions in integers, and may have none when $n =$ or > 3; Example when $n=4$ $A=5$ and $B=1$.

10: $p-qx$ is less, abstracting from its sign, than it would be if we substituted for p and q any smaller integers, when $\frac{p}{q}$ is a convergent to x; *Lagrange's* elegant proof of the converse theorem; the value of a periodic CF when continued to inf^y. whether it begins with periodic terms or not, is always the root of quadratic equation having rational coefts. Easy method of finding this quadratic.

11.—The root of *any* quadratic equation whose coefts. are integers demonstrated to be a periodic C F, easy method of ascertaining a priori the very point at which the periodicity begins: the 1st or 2nd denr. of the C F for $\sqrt{\frac{A}{B}}$ begins the period according as A > or < B; method of finding the length of the period of the C F for $\frac{B+\sqrt{C}}{A}$ when only C is given; this *capital point* was mentioned, but intentionally omitted by *Lagrange*: example worked at full length.

12. Method of finding the C F for the root of a quadratic equation ; useful contraction of the process ; the commencement of the periods known by simple inspection ; previous reduction of $\frac{B+\sqrt{C}}{A}$ sometimes necessary in order to convert it into a C F ; Examples ; the C F[s] for the square roots of certain integers found very simply ; the period of C. F. for $\sqrt{\frac{A}{B}}$ is symmetrical with the exception of its last term which is always twice the greatest integer in $\sqrt{\frac{A}{B}}$; in converting $\frac{B+\sqrt{C}}{A}$ into a C F if the denr. A of any *complete* quotient be $=1$; the corresponding term B of the numerator will be the integer next $<\sqrt{C}$.

13 —Very expeditious method of finding the min^{m}. value of $Ap^2+Bpq+Cq^2$; when p and q are unknown integers, without the trouble of previously finding p and q ; rule to know a priori the sign ($\pm$) of the min^{m}. value ; when the number of terms in the period of the C. F. for either root of $Ax^2+Bx+C=0$ is *odd*, both $+E$ and $-E$ will then be possible values of $Ap^2+Bpq+Cq^2$, E being the denr. of any one of the *complete periodic* quotients of the C F for x ; Example fully explained.

14 : $x^2-Ny^2=1$ is always possible in integers, when the given integer (or fraction) N is not a square ; $x^2-Ny^n=-1$ possible in integers only when N is the sum of two integral squares prime to each other ; *when* $x^2-Ny^2=-1$ *is possible it admits of a solution in integers lower than the least integers* (*except* 1 *and* 0) *that fulfil* $x^2-Ny^2=1$; $x^2-Ny^2=\pm H$ is always impossible in integers when H is $<\sqrt{N}$ and not found among the denoms. of the complete quotients of C. F. for $\sqrt{N}$; Examples.

15.—*All* the solutions in integers of $x^2-44y^2=4$, $x^2-44y^2=-8$, $x^2-61y^2=\pm5$ obtained by means of continued fractions, several useful remarks ; *Lagrange's* method of deducing the solution of $x^2-Ny^2=\pm H$ when $H>\sqrt{N}$ from the solution of $x^2-Ny^2=\pm H'$ when $H'<\sqrt{N}$; examples $x^2-23y^2=-7$; $x^2-13y^2=-9$; impossibility of $79y^2+101=\square$ in integers.

16.—Impossibility of a solution in integers of $x^2-Ny^2=\pm H$, 1° when $N=a^2+1$ and H (not a square number) is >1 and $<\sqrt{N}$: 2° when $N=a^2-1$ and H (not a square) is >1 and $<a$, this case also impossible if $H=-1$; 3° when $N=a^2+a$ and H (not a square) is >1 and $<a$ or when $H=-1$ or $+a$; 4° when $N=a^2-a$ and H (not a square) is >1 and $<a-1$.

17.—How to find solutions in integers *ad lib.* of $x^2-Ny^2=1$ from one known solution of it : when $N=a^2\pm\frac{2a}{b}$ then $y=b$ is one known solution ; from one known solution of $x^2-Ny^2=-1$ we can find solutions in integers *ad lib.* of each of the equations $x^2-Ny^2=\pm1$; By means of the auxiliary equation $x^2-Ny^2=1$ we can find solutions in integers *ad lib.* either of $x^2-Ny^2=A$ or of $ax^2+bx+c=\square$ from one known solution of either.

18.—*Lagrange's* method (*much simplified*) for finding at once in terms of n the nth convergent to the CF for the root of a quadratic without the trouble of previously calculating all the preceding convergents. Example explained at full length.

NAMES OF SUBSCRIBERS.

Note.—The numerical figures placed after the names of some Subscribers, indicate the number of copies Subscribed for, and one copy only was subscribed for by those whose names are followed by no figures.

Abbott, Thomas K., Esq., M.A., F.T.C.D.
Anderson, James, Esq., Seghill, Northumberland
Andrew, Rev. James, M.A., Dublin
Andrews, Rev. Mr. B.A., Kilkenny
Anster, John, Esq., LL.D., Regius Professor of Civil Law, T.C.D.
Apjohn, James., Esq., M.D., Professor of Chemistry and Minerology, T.C.D. 5
Arnold, H., Esq., Training Inst., Bandon

Babington, Thomas, Esq., Attorney, Cork
Banim Michl., Esq., Postmaster, Kilkenny
Barlow, Rev. James Wm., M.A., F.T.C.D.
Barrett, Mr. Rd. Newcastle, Co. Limerick
Barry, Chas. R., Esq. M.A., Barrister-at-Law, Dublin 5
—— David, Esq., B.A. Moderator, T.C.D. Kingstown
Bell, David Charles, Esq.. Dublin
Benson, Sir John, Cork. 3
Blakiston, Matthew, Esq., C.E., Dublin
Blunden, Sir John, Bart., Castleblunden, Co. Kilkenny, 2
—— Wm. P, Esq., B.A., Co. Kilkenny 2
Bolger, Mr. James, Dublin
Booth, John L., Esq., B.A., Pococke Institution, Co. Kilkenny 2
Bower, John, Esq., C.E., Dublin
Bowhen, Mr. James, Newcastle, Co. Lim.
Boyd, Mr. Joseph, Prof of Chem. Kilkenny
Boyle, Swithin, Esq., Student, T.C.D.
Bradford, Mr., Sch. T.C.D.
Brady, Rev. Thos., Seneca-falls, State of New York
Bradley, Thos. Esq., Richview, Kilkenny
Brazill, Thomas, Esq , C.E., Dublin
Brett, H., Esq., C.E. Co. Surveyor, Dublin 2
Brown, Stephen, Esq., LL.D., Bandon 2
—— Rev. John, LL.D., Kilkenny 2
—— Valentine, Esq., C.E., Dublin
Brookes, Charles H., Esq , C.E., Newcastle-upon-Tyne
Buckley, Eugene, Esq., Hawarden, near Chester
—— Mr. Patrick, Bonmahon, Co. Waterfd.
Burke W. J. Esq. C.E., Dublin
—— Lawrence, Esq., Land Surveyor, Dub.
—— John Lardner, Esq., LL.D., Dublin 2
Bunbury, Benjamin, Esq., Kilkenny
Burgess, Wm. Charles, Esq., Shannon Foundry, Limerick
—— Wm. Esq., Architect & Builder, Lim.
Burnes, Mr Jas., St. Mary Shandon Sch. Ck.
Butcher, Rev. Samuel, D.D., Ex-F.T.C.D. Professor of Divinity, T.C.D.
Buttery, John, H.M., Dockyard, Chatham
Butler, Eward, Esq., M.A., Professor of Mathematics, Catholic University 2

Cameron, Mr. William, Dublin
Campbell, Owen, Esq., Southampton
Campion, John Thomas, Esq., Kilkenny
Cane, Robert, Esq , M.D., Kilkenny
—— Robert E., Esq., A.B., M.B. Lieut. Royal Artillery
Carmody, Thomas, Esq., Castlequarter, near Tarbert
Carson, Rev. Joseph. D.D., F.T.C D.
Casey, Mr. J., Head Mast. Model Sch Kil.
Chalmers, J.M , Esq., Sunderland
Chapman, Wm., Esq., Monkwearmouth
Checkley, Francis L., Esq., Sizar, T.C.D.
Chetwode, Edward Wilmot, Esq., M.A., M.R.I.A., Woodbrook, Queen's Co. 6
—— Knightley Wilmot, Esq., B.A. Moderator, T.C.D. 6
Clements, Rd. Esq., Up. Islington-st., Lond.
Cockle, James, Esq., M.A., F.R.A.S. F.C.P.S., Hyde Park, London
Collins, John, Esq., Ex-Sizar, T.C.D., 13 Anglesea-street, Dublin 5
—— Francis, Esq., 16th Lancers
Conner, Rev. R. Mountifort. M.A, F T.C.D.
Connors, Mr. Terence, Glenquin, near Newcastle, Co. Limerick
Conwill, Mr. John, Leighlin Bridge
Corcoran, Mr. Jn. Loriga Mills, Skibbereen
Costello, Ed., A.B. T.C.D., Mnkstn. Cresent
Coyne, John, Esq., Kilkenny 2
Crofton, M. P., Esq., Dublin
Cromie, Henry, Esq., Student T.C.D.
Cross, M., Esq, Natl. Bd. Educn. Dub. 2
Cullen, Daniel, Esq., J.P., Kilkenny, 3
Cully, Rd., Esq., Bank of Irld. office, Kil.
Cumming, James H., Gd. Parade, Cork 2
Curry, Eug., M.R.I.A., Prof. in Cath. Univ.
Curtin, Mr. Laur., Tournafulla, Co. Lim.
Cussen, John, Esq. Ardagh, Co. Lim. 3

Daft, T. Esq., C.E. Seville place, Dublin
Davis, Mr. D., woolen manuf., Kilkenny
Davison, James, Esq., Sunderland
Dickinson, James A. Esq., C.E., Dublin
Dickson, Rev. B., M.A., F.T.C.D.
Dodd, J. P. Esq., LL.D., North Shields
Dillon, J. B. Esq., B.A. Barrister-at-law, Dublin
Dobson, Thomas Esq., B.A. Royal Hospital, Greenwich
Dockeray, John, Esq., Santry, Dublin 3
Doheney, Wm. Esq., C.E., Dublin

Donelly, Mr. P. J. land surveyor, Dublin
Donovan, M. Liscordan, Ardagh Co. Lim.
—— Jos. Esq., Grouse Lodge, Co. Lim.
Douglas, David, G. Esq., LL.D., Cork 2
—— W. J. Esq., Kilkenny 2
Dowden, Richard, Esq., Cork
Drapes, Rev. John L., M.A., Kilkenny
—— Rev. Vernon, Do. Do.
Duckett, Robert, Esq., Dublin
Duffy, Henry, M. Esq., C.E., Dublin
Dunbar, Wm. Esq., M.A. Ex-Sch. T.C.D. Talbot-place, Dublin
Dundas, Rev. W. J., Dublin

Eager, Charles, H. Esq., Dublin
Edge, Rev. John, A.B., Ex-Sizar, T.C.D.
Empson, John, Esq., C.E., Kilkenny
Engledow W. H. Esq, Pococke Institution, Kilkenny 2
English, John, Esq., C.E., Dublin

Fagan, William, Esq., M.P., Cork 2
Fawcett, Mr. Joseph, 2 Chester-road Sunderland
Falvey, Mr. Daniel W., Bandon 2
Feely, John, Esq, B.A., Bandon
Ferguson, Daniel, Esq., Kilkenny
Fitzgerald, Gerald, Esq., M.A., M.D., Portabello, Dublin
—— John, Esq., M.A., Kingstown
—— James, Esq., eminent writer and accountant, Dublin 2
—— Edward, Esq., Tipperary
Fitzgibbon, Rt. Worshipful Wm. Mayor of Cork 2
Fitzpatrick, Ds. Esq., Science Sch. T.C.D.
Fitzsimons, John, P. Esq., Medical Hall Kilkenny.
Fogarty, Rev. D., Clarendon-st., Dublin
Foley, Rev. Daniel., M.A. Professor of Irish T.C.D.
Frizelle, Rev. Richard, B.A., Dublin

Gabbett, Rev. Robt. J., M.A Shanagolden
Galbraith, Rev Joseph, M.A., F.T.C.D. Professor of Ex. Phys. T.C.D.
Galvin, John, Esq., B.A. Ex-Sch. T.C.D. translator of Faust, &c.
Ganley, Patrick, Esq, C.E., Dublin
Gartland, William, Esq., C.E. and land surveyor, Rathmines
Gerty, Rev. Hamilton, B.A., T.C.D.
Gibson, Thomas, Esq., Chester-road, Sunderland.
Goninan, Wm. Esq., Endowed School, Chevely, Cambridgeshire 2
Gray, Nich. Esq., Tinney Park, Co. Kilkenny
—— Wm. Esq, Heytesbury-st., Dublin
Graves, Rev Charles, D.D., F.T.C.D., Professor of Mathematics, T.C.D.
—— John T. Esq, M.A., Cheltenham 2
Gorman, Rev. W. C., B.A., Kilkenny 2
Gowdy, Rev. R. T., M.A., Cork
Greene, John B. Esq., C.E., Dublin
Gordon, Sir John, Cork
Griffith, Rich. Esq., LL.D., Dublin 2
Gumley, Rev. John B. A. Glebe House, Fuerty, Co. Roscommon 2
Gwynn, John, Esq., M.A., F.T.C.D.

Hackett, Sir William, Cork
Hamilton, Sir William R., LL.D., F.R.A.S. Andrews, Professor of Astronomy, T.C.D., Astronomer Royal of Ireland, &c. &c. 5
—— John, Esq., Cove-street, Cork
Hancock, Wm. Neilson, Esq., LL.D. Professor Q.C. Belfast 4
Harcourt, Joseph, Esq., Newry 4
Hare, Mtths, Esq., LL.D. Step.-green, 2
Hargreave, Chas. Js. Esq., LL.D., Dub. 3
Harte. Andrew S. Esq., LL.D., S.F.T.C.D.
Haughton, Rev. Samuel, M.A., F.T.C.D., Professor of Geology, T C.D.
Hayden, Wm. Esq., Solicitor, Kilkenny
Healy, Michl., Esq., B.A., Ex-Sizar, T.C.D.
Heffernan, D. G. Esq., C. E., Dublin
Henessy, Rev. J. St. Kyran's College, Kilkenny
Hemans, George W. Esq., C.E., Dublin 2
Hewson, Thomas Esq., T.C., Railway Hotel, Kilkenny
Higgin, Right Rev. Wm., D.D., Lord Bishop of Derry and Raphoe 4
Hobson, Rev. John M., B.A. Kilkenny
Hodnett, W.P., Esq., Kilkenny, Fusiliers
Hogan, Mr. Michl., Nat. School, Gowran
Holt, Edward Grattan, Esq., C.E., A.B., T.C.D., Celbridge
Howard, John F., Esq., B.A., Barrister-at-Law, Dublin 2
Hughes, Wm., Esq., Summer-hill, Dub.

Ingram, John Kells, Esq., LL.D., F.T.C.D. Professor of Oratory, T.C.D.

Jacob, Wm. Henry, Esq., B.A. Portarlington 2
James, John, Esq., M.D. Kilkenny
Jeffares, W. S. Esq., Bank of Ireland Office, Limerick
Jellett, Rev. John H., F.T.C.D., Prof. of Natural Philosophy, T.C.D.
Johnson, Alex. Esq., B.A. senr. Moderator, T.C.D.
—— James, Esq., Dublin
—— Edward, Esq., Bandon
Jones, Philip, Esq., M.R.I.A. Nutgrove, Rathfarnham 2
—— Rev. Abraham, M.A., Rathmines
Jordan, Edmund, Esq., M.A., Barrister-at-Law, Dublin 2
Joynt, William L. Esq., Limerick

Kane, Sir Robert, F.R.S. President of Queen's College, Cork 5
Kavanagh, P. Esq., M.D. Kilkenny
—— James, Esq., Stephen-st., Dublin
—— James, W., Esq., Rathgar, Dublin, 4
Kearns, John, Esq., L.R.C.S I. Kilkenny
Kelly, Michael, Esq., Knockmahon, Co. Waterford
Kennedy Mr. John, Ballybunion, Co Kerry
Kennelly, Mr. Maurice, Newcastle, Co. Limerick
Kenny, Patrick, Esq., Butter Merchant, Kilkenny
Kerr, Wm. P. Esq., B.A. Inspector of Schools 2
Kinchella, Louis, Esq. A.B., M.D., Kilkenny
Kincella, Mr. Patrick, Leighlin Bridge

Lalor, Joseph, Esq., M.D., Lunatic Asylum, Kilkenny
Lanigan, Rt. Worshipful Wm., Mayor of Kilkenny
Lardner, Dionysius, Esq., LL.D., Paris
Larminie, Wm. Rea, Esq., Sizar, T.C.D.
Lee, Rev. Wm., M.A., F.T.C.D.
Leech, Arthur Blenerhasset Esq. Collooney
Lefanu, Wm., Esq., C.E., Fitzwilliam square, Dublin
Leeper, Rev. Alexr., M.A., M.R.I.A. Dublin
Leeson, James, Esq., C.E., Land Valuator, Booterstown, 2
Lloyd, Rev. Humphrey, D.D., S.F.T.C.D. 3
Lindwart, Percy, Esq., Ph. D., Kilkenny 2
Locke, John, Esq., M.A., Auction Clerk of Incumbered Estates 2
Lonergan, Mr. Patrick, Builder, Kilkenny
Longfield, Montifort, Esq., L.L.D., Q.C., Commsr. of Incumbered Estates, &c.
——Rev. George, M.A., F.T.C.D.
Luby, Rev. Thomas, D.D., S.F.T.C.D. 2
Lynch, Mr. Rodolph Banogue, Croom, County Limerick

Macdonnell, The Rt. Hon. Alexander Resident Commsr. of Natl. Educn. 6
Macdonnell, Rev. Richard, D.D., Provost of T.C.D. 3
M'Donald, Rev. Edward, D.D., St Kyran's College, Kilkenny
M'Cabe, Mr. John, Carrickmacross
M'Carthy, Mr. Denis, Marine School, Waterford
Maguire, John F., Esq., M.P., Cork 2
M'Guire, Edward, Esq., B.A., M.B., Galway
M'Cormick, Edward, Esq., C.E., Land and Engineering Surveyor, Grosmont, near Hereford 7
——Rev. William, C.C., Moate, County Westmeath 2
M'Creanor, Edward, Esq., Inspector of National Schools
M'Curdy, John, Esq., Architect, T.C.D.
M'Donagh, Rev. Charles, M.A., Enniskerry
M'Dowell, George, Esq., M A., F.T.C.D.
M'Elligott, Denis, Esq., C.E., and Land Surveyor
M'Elroy, Mr. James, Waterford
M'Eniery, James, Esq., Newcastle, County Limerick
Macgrath, Joseph, Esq., Killkenny
M'Kean, Andrew, Esq, Provincial Bank, Kilkenny
M'Kittrick, George, Esq., C.E., Shirley Office, Carrickmacross
M'Mullen, James A., Esq., A.B., Douglas, Isle of Man 2
M'Namara, Timothy, Esq., Mathematician, Ballaghderrin, Co. Mayo
M'Neece, Rev. Thomas, D.D., ex F.T.C.D
M'Neill, Sir John. Mount Pleasant, Co. Louth 6
Malett, Rev. John A., M.A., F.T.C.D., Numismatist, T.C.D.
Malone, Charles, Esq., Barrister-at-law, Dublin.
Manso, Herr E. Theodore, Tivoli Hall, Kinstown
Marani, Signor, Professor of Spanish and Italian, T.C.D.,
Martin, Mr. James, Pennefeather's lane, Kilkenny
Merry, Mr. Patrick, Kilkenny
Miller, George M., Esq., C.E., G. S. and W. R., Dublin
—— Charles, Esq., Registrar of Electors, T.C.D.
—— W. J., Esq., B.A., Eltham, Kent
Milne, J., Esq., LL.D., Dolar Institution, Clackmannanshire
Mitchell, Joseph A., Esq., Short Heath, Willenhall, Staffordshire 2
Moffet, Thomas, Esq., LL D., Professor Q.C., Galway
Moore, Patrick, Esq., Raheny 2
—— Rev. John Lewis, D.D., S.F.T.C.D.
—— Mr. Michael Newcastle, County Limerick
Moorhead, Mr. Wm., Hibernian School, Phœnix Park
Moran, Owen, Esq., Senecca Falls, State of New York, formerly of Carrickmacross
Mortimer, William, Esq., A.B, senior Moderator, T.C.D.
Mulvany, Christ., Esq., C.E., Dublin
Murphy, Hugh, Esq., Killyshandra, Co. Cavan 3.

Nash, R. W. H., Esq., Drumcondra, Dublin
—— Rev. John, Kilkenny
Neville, John, Esq., C.E., M.R.I.A., County Surveyor, Dundalk 20

—— Parke, Esq., Do. Do., Dublin
Newell, F. W., Esq., M.A., Cork
Nicholson, William, Esq., Kilkenny
North, Rev. Roger, M.A., Rathmines

O'Brien, Rt. Rev. James Thomas, D.D., Lord Bishop of Ossory and Ferns 6
O'Brien, William Smith, Esq., Cahermoyle, County Limerick 7
—— Rev. Patrick, C.C., Thomastown, Co. Kilkenny 2
—— The Hon. Justice, Dublin
O'Brennan, Martin A., Esq., LL.D., Bolton street, Dublin
O'Callaghan, Mr. James, Mathematician, Clogheen
O'Carroll, Rev. J., B.A , Ex Sch., T.C.D.
O'Donnell, Mr. Patk., Nat. Sch. Cappamore
O'Donovan, John, Esq. LL.D., M.R.I.A., Professor Q.C., Belfast
O'Farrell, James, Esq., Southampton
O'Flaherty, Martin, Esq., B.A., Solicitor, Dublin and Galway 2
O'Herlihy, John, Esq., St. Laurence, Seminary, Dublin
O'Mahony, Rev, Thaddaeus, Senr. Ethical Moderator, T.C.D.
O'Neill, John, Esq., Cork
O'Regan, Mr. James, Ballina
—— Rev. John, M.A., Rector of Finglass
O'Reilly, Edward Fletcher, Esq., Dublin
Ormonde. the Most Noble, the Marchioness of, Kilkenny Castle 6
O'Shaughnessy, Mr. Laurence, Newcastle, Co. Limerick
O'Sullivan, Daniel, Esq., Cork

Piggott, Lieutenant, 16th Lancers
Poole, Rev. Hewit R., M.A., F.T.C.D.
—— Thomas, Walter Esq., Dublin 2
Porter, Rev. Jn. Leech, B.A. Modr., T.C.D. 2
—— Rev. James, Manse, Kilkenny
Potter, John, Esq., M.B., Kilkenny 2
Power, Edward, R. Esq., B.A., Harrington-street, Dublin
Price, Rev. Wm. J., LL.D., Waterford
Prim, John, G. A. Esq., Editor and Proprietor of "the Kilkenny Moderator"
Purtill, Patrick, Esq., Dublin 2

Quarman, Miss Elizabeth, Castlemartyr Co. Cork 2

Ralph, C. S. Esq., B.A. Barrister-at-law, Dublin 2
Ranalow, William, Esq., Kilkenny
Ringwood, Rev. Frederick H., M.A. Royal School, Dungannon
Roberts, Rev. William, M.A., F.T.C.D.
—— Michael, Esq., Do. Do.
Robinson, John Joshua Esq., H.M. Dockyard, Portsea
Rogan, Anthony, Esq., Southhampton
Rudkin, W. H. Esq, Up. Baggot-st., Dub.
Rutherford, Wm. Esq., LL.D., F.R.A.S. Professor R.M A., Woolwich
—— Major, Dublin
Rutledge, Rev. John Young, M.A. Ex-F.T.C.D. 2
Rutter, Edward, Esq., Mathematician, Sunderland, 2
Ryan, Mr. Laurence, Mathematician, Leighlin Bridge
—— Mr. John, Francis Do., Kilkenny
—— Mr, Thomas, Grocer, Kilkenny

Sadleir, Rev Wm. Digby, D.D., S.F.T.C.D.
Salmon, Rev. George, M.A., F.T.C.D.
Scriven, E.H. Esq., M.D. L.R.C.S. Roundtn.
Scott, Rev. Thomas, M.A., Rathmines 2
Seymour, Joseph, Esq., Patrick-st. Cork
Shaw, George, F. Esq , LL.D., F.T.C.D.
—— Thos. S. Esq., B.A., Carrickmacross
Sheehan, Timothy, Esq., M.A., Head Inspector of National Schools 2
Sheppard, Wm. Esq., Bradford, Yorkshire
Sheehy, Henry, Esq., Ballingarry, Co. Limerick
—— John, Esq., Ardbohill Rathkeale, Co. Limerick
—— Edward, Esq., Inspector of National Schools, 2
Shortall, Michl. Esq., Solicitor, Kilkenny
Shyne, Bartholomew Esq., Dublin
Sloane, John, S. Esq., C.E., Dublin
—— George, Esq., C.E. and Land Surveyor, Shercock, Co. Cavan
Smith, Beresford, B. Esq., Dublin
—— Frederick, Esq., A.B. Senr. Modr., T.C.D.
—— H. Ambrose Esq., Northern Assurance, Co. Aberdeen
—— John, Esq., National Bank, London 2
Stackpoole, Rev. Wm. C., LL.D., Kingstn.
Stapleton, Wm. Esq., B.A., Dublin
Staunton, Michael, Esq., Collector General of Taxes, Dublin
Stevenson, John, Esq., Alliance Gas Company, Dublin
Steele, Rev. Wm., M.A., Endowed School Enniskillen 3
Stitt. Rev. George, M.A., Modr., T.C.D.
St. John, James, Esq., LL.D., Kilkenny 2
Stubbs, Rev., John Wm., M.A., F.T.C D.
Sullivan, Michael, Esq,, M.P., Kilkenny 6
—— Francis, Esq., Kilkenny
—— Robert, Esq., LL.D., Professor at Bd. of National Education, Dublin 5
—— Daniel, Esq., B.A., Dublin
—— W. K. Esq., Ph. D. Professor of Chemistry, Dublin
Swanton, Rev. Mr., Kilkenny

Taylor, Edward, Esq., Kildare-place, Dublin 2
—— Wm. Esq., Mechanics Inst ,Liverpool
—— John, P. Esq., Sch, T.C.D.
Tanner, William, Esq., M.D., Cork
Thompson, Mr. Peter, Renhaw-street, Manchester

Tidmarsh, Jms. M. Esq., J.P., Kilkenny 3
Tolekcn, John, Esq , MD. S.F.T.C.D.
Townsend, Rev. Richard, M.A., F.T.C.D.
Traynor, James, Esq., C.E. and Land Surveyor, Shercock Carrickmacross 2
Troy, Wm. De Esq , Hawarden Chester
Turner, John, W. Esq., Educational Institute, Dundalk 4

Underwood, Thomas, O. Esq., Earlsfort Terrace, Dublin
Uncles, Mr. Joseph, Bandon

Walker, Richard Esq., Mathematician, Newcastle, Co. Limerick.
—— Alexander, Esq., Carrickfergus
Wall, Rev. Chars. Wm., D.D., S.F T.C.D
—— Rev. Richard H., D.D., Dublin
Walsh, Richard, Esq., M.A., T.C.D.
Welch, John, Esq , Monkwearmouth
Washington, Mr. George, Kilkenny
Watson, Stephen, Esq. Haydon Bridge, Northumberland
Watters, P. Esq , Solicitor, Kilkenny
Webster, Rev. George, M.A., Donnybrook
Welstead, J. H. Esq., B.A., Cork
Wesleyan.Connexional School hd.master of
Whittaker, Rev. Thos., B.A., Kingstown
Whitty, John, Irwin, Esq ,LL.D., Civil and Mining Engineer, Dublin
Wickham, Mr J.D., Kilkenny
Wiley, William, Esq., LL.D. Senior Moderator, T.C.D.
Wilkinson, T. T. Esq., F R.A.S., Burnley, Lancashire
Williamson, Benj. Esq., M.A., F.T.C.D.
Wilson, Rt. Rev. James, D.D., Lord Bishop of Cork &c. 6
—— Rev. Richard, D.D., Chelsea, London
Woolhouse, W. S. B. Esq., F.R.A.S. &c. &c., London
Worrall, John, L. Esq., C.E., Booterstown
Woulfe, James, Esq., Dublin

ADDITIONAL SUBSCRIBERS.

Samuel Adair, Esq., Insp. of Nat. Schs.
Eugene A. Conwell, Esq., Do.
Michael Coyle. Esq., Do., 2

ERRATA.

Page 3	*line* 34	Equations	*should be*	equation.
4	—— 13	primitive	——	prime
5	*last line but* 1 ;	bofore	——	before
7	*line* 12	$p^2.n$	——	p^2,n
7	—— 12	mn	——	$m-n$
11	*last line but* 7 ;	itself sum	——	itself the sum
13	—— 14	z^2	——	z'^2
15	—— 14	and p	——	nor p
17	—— 6	$^2+y^2$	——	x^2+y^2
18	—— 6	ay	——	ay^2
21	—— 33	ond	——	and
25	—— 38	not	——	not
29	—— 17	$8,1^0$	——	by art $8,1^0$
——	—— 21	ond	——	and
36	*last line,*	proportion	——	proposition
49	*line* 4	forn	——	form
49	—— 31	Hence	——	Hence
52	—— 26	simultaneous	——	simultaneous Equations

Spotted something that isn't quite right?

Let us know and we'll fix it for future reprints. And as a thank you for your diligence, we'll send you a free Sothis book of your choice.

publisher@sothis.press

This page intentionally left blank

Except of course for that line above. And this one.
And whilst we have your attention, perhaps you'd
like to follow us on Instagram @sothis.press. Go on.

Made in United States
North Haven, CT
04 November 2026

21766931R00044